4 page

no eccentricity

no domain of function

$$y = \frac{x}{x^2 + 1}$$

ANALYTIC GEOMETRY

FOURTH EDITION

GORDON FULLER
Professor Emeritus of Mathematics
Texas Tech University

ADDISON-WESLEY PUBLISHING COMPANY

Reading, Massachusetts · Menlo Park, California
London · Amsterdam · Don Mills, Ontario · Sydney

This book is in the

ADDISON-WESLEY SERIES IN MATHEMATICS

ISBN 0–201–02102–1
IJKL–MA–798

Preface

This fourth edition of *Analytic Geometry*, like the earlier editions, has been carefully tailored for a first course in the study of the subject. It emphasizes, especially, the basic concepts which are needed in calculus and in many other areas of mathematics. The organization, however, has been changed in order to obtain a more logical and cohesive treatment, and the exposition has been carefully upgraded. Also, new material has been included in several places. In particular, we have introduced lucid discussions of Relations and Functions, Families of Circles, and Surfaces of Revolution, and we have extended the treatment of Direction Cosines and Direction Numbers of lines.

Teachers of the earlier editions will observe that all of the chapters of the present edition have been significantly strengthened. More especially is this true of the following chapters:

2. The Straight Line and the Circle
4. Simplification of Equations
5. Polar Coordinates
9. Vectors, Planes, and Lines

The study of conics has been deservingly treated more fully than is essential for the study of calculus. If desired, however, the coverage of this part can be shortened by assigning fewer problems from the exercises. The translation of axes, used earlier with the circle, has been applied in passing from the simplified equations of conics to the more general forms involving h and k.

In Chapters 8 and 9 the elements of solid analytic geometry have been treated. The first of these chapters deals with quadric surfaces, and the second one deals with planes and lines. This order was chosen because a class which takes only one of the two chapters should, preferably, study space illustrations of second-degree equations. Vectors are introduced and applied in the discussion of planes and lines. This study has been facilitated, of course, by the use of vectors, and it also provides the student with more than a passing encounter with this valuable concept.

The solved problems bear a close correlation with the problems found in the exercise sections. This feature, together with the careful exposition, tends to give the book a self-teaching quality. The exercise sections occur at short intervals, and each contains an abundance of problems. Also, numerous problems of theoretical implication have been included. These, of course, should afford a definite challenge to some of the students. Five numerical tables in the Appendix should meet any needs which may arise in solving the problems. Answers to the odd-numbered exercises are bound with the book, and answers to the even-numbered exercises are available in a separate booklet.

This book is intended as a semester course for senior high-school students or for freshmen in colleges and universities. Students in high school with four or five class meetings a week should cover most, if not all, of the book. Students in college, however, where three meetings a week are more likely, will in some instances not be able to cover the entire book. If the entire book is not covered, either in high school or college, we suggest that Sections 4 and 5 of Chapter 6, Section 9 of Chapter 9, and all of Chapter 10 be made the primary candidates for omission.

We are indebted to the Van Nostrand Reinhold Company for their permission to use certain material from their book *Analytic Geometry and Calculus*, 1964, by Gordon Fuller and Robert M. Parker. Also the McGraw-Hill Book Company has kindly agreed to our use, in Chapter 1, of some material from their book *Plane Trigonometry with Tables*, Fourth Edition, 1972, by Gordon Fuller.

Lubbock, Texas   G. F.
November 1972

Contents

Chapter 4   Simplification of Equations

Chapter 5   Polar Coordinates

Chapter 6   Parametric Equations

Chapter 7   Algebraic Curves

Chapter 1

Fundamental Concepts

For several centuries geometry and algebra developed slowly, bit by bit, as distinct mathematical disciplines. In 1637, however, a French mathematician and philosopher, René Descartes, published his *La Géométrie*, which introduced a device for unifying these two branches of mathematics. The basic feature of this new process, now called **analytic geometry**, is the use of a coordinate system. By means of coordinate systems algebraic methods can be applied powerfully in the study of geometry, and perhaps of still greater importance is the advantage accruing to algebra by the graphical representation of algebraic equations. Indeed, Descartes' remarkable contribution paved the way for rapid and far-reaching developments in mathematics, and analytic geometry continues to be of great value in mathematical investigations.

1. DIRECTED LINES AND SEGMENTS

A line on which one direction is chosen as positive and the opposite direction as negative is called a **directed line**. A segment of the line, consisting of any two points and the part between, is called a **directed line segment**. In Fig. 1, the positive direction is indicated by the arrowhead. The points A and B determine a segment, which we denote by AB or BA. We specify that the distance from A to B, measured in the positive direction, is positive; and the distance from B to A, measured in the negative direction, is negative. These two distances, which we denote by $\overrightarrow{AB}$ and $\overrightarrow{BA}$, are called **directed distances**. If the length of the line segment is 3, then $\overrightarrow{AB} = 3$, and $\overrightarrow{BA} = -3$. Distances, therefore, on a directed line segment satisfy the equation

$$\boxed{\overrightarrow{AB} = -\overrightarrow{BA}}$$

Another concept with respect to distance on the segment AB is that of the **undirected distances** between A and B. The undirected distance is the length of the segment, which we take as positive. We will use the notation $|AB|$ or $|BA|$ to indicate the positive measurement of distance between A and B, or the length of the line segment AB.

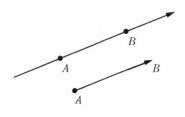

Figure 1

In view of the preceding discussion we may write

$$\overrightarrow{AB} = |AB| = |BA| = 3$$
$$\overrightarrow{BA} = -|AB| = -|BA| = -\mathbf{3}$$

Frequently the concept of the absolute value of a number is of particular significance. Relative to this concept, we have the following definition.

DEFINITION 1   *The **absolute value** of a real number a, denoted by $|a|$ is the real number such that*

$$|a| = a \text{ when a is positive or zero}$$
$$|a| = -a \text{ when a is negative.}$$

According to this definition, the absolute value of every nonzero number is positive and the absolute value of zero is zero. Thus,

$$|5| = 5 \qquad |-5| = -(-5) = 5 \qquad |0| = 0$$

THEOREM 1   *If A, B, and C are three points of a directed line, then the directed distances determined by these points satisfy the equations*

$$\overrightarrow{AB} + \overrightarrow{BC} = \overrightarrow{AC} \qquad \overrightarrow{AC} + \overrightarrow{CB} = \overrightarrow{AB} \qquad \overrightarrow{BA} + \overrightarrow{AC} = \overrightarrow{BC}$$

Proof. If B is between A and C, the distances $\overrightarrow{AB}$, $\overrightarrow{BC}$, and $\overrightarrow{AC}$ all have the same sign, and $\overrightarrow{AC}$ is obviously equal to the sum of the other two (Fig. 2). The second

Figure 2

and third equations follow readily from the first. To establish the second equation, we add $-\overrightarrow{BC}$ to both sides of the first equation and then use the condition that $-\overrightarrow{BC} = \overrightarrow{CB}$. Thus,

$$\overrightarrow{AB} = \overrightarrow{AC} - \overrightarrow{BC} = \overrightarrow{AC} + \overrightarrow{CB}$$

2. THE REAL NUMBER AXIS

A basic concept of analytic geometry is the representation of all real numbers by points on a directed line. The real numbers, we note, consist of the positive numbers, the negative numbers, and zero.

To establish the desired representation, we first choose a direction on a line as positive (to the right in Fig. 3) and select a point O of the line, which

Figure 3

we call the **origin**, to represent the number zero. Next we mark points at distances 1, 2, 3, and so on, units to the right of the origin. We let the points thus located represent the numbers 1, 2, 3, and so on. In the same way we locate points to the left of the origin to represent the numbers $-1, -2, -3$, and so on. We now have points assigned to the positive integers, the negative integers, and the integer zero. Numbers whose values are between two consecutive integers have their corresponding points between the points associated with those integers. Thus the number $2\frac{1}{4}$ corresponds to the point $2\frac{1}{4}$ units to the right of the origin. And, in general, any positive number p is represented by the point p units to the right of the origin, and a negative number $-n$ is represented by the point n units to the left of the origin. Further, we assume that every real number corresponds to one point on the line and, conversely, every point on the line corresponds to one real number. This relation of the set of real numbers and the set of points on a directed line is called a **one-to-one correspondence**.

The directed line of Fig. 3, with its points corresponding to real numbers, is called a **real number axis** or a **real number scale**. The number corresponding to a point on the axis is called the **coordinate** of the point. Since the positive numbers correspond to points in the chosen positive direction from the origin and the negative numbers correspond to points in the opposite or negative direction from the origin, we shall consider the coordinates of points on a number axis to be **directed distances** from the origin. For convenience, we shall sometimes speak of a point as being a number, and vice versa. For example, we may say "the point 5" when we mean "the number 5," and "the number 5" when we mean "the point 5."

3. RECTANGULAR COORDINATES

Having obtained a one-to-one correspondence between the points on a line and the system of real numbers, we next develop a scheme for putting the points of a plane into a one-to-one correspondence with a set of ordered pairs of real numbers.

DEFINITION 2   *A pair of numbers (x, y) in which x is the first number and y the second number is called an* **ordered pair**.

We draw a horizontal line and a vertical line meeting at the origin O (Fig. 4. The horizontal line OX is called the **x axis** and the vertical line OY, the

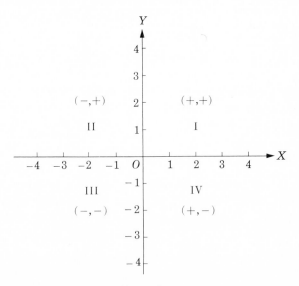

Figure 4

y axis. The x axis and the y axis, taken together, are called the **coordinate axes**, and the plane determined by the coordinate axes is called the **coordinate plane**. The x axis, usually drawn horizontally, is called the **horizontal** axis and the y axis the **vertical** axis. With a convenient unit of length, we make a real number scale on each coordinate axis, letting the origin be the zero point. The positive direction is chosen to the right on the x axis and upward on the y axis, as indicated by the arrowheads in the figure.

If P is a point on the coordinate plane, we define the distances of the point from the coordinate axes to be **directed distances**. That is, the distance from the y axis is positive if P is to the right of the y axis and negative if P is to the left, and the distance from the x axis is positive if P is above the x axis and negative if P is below the x axis. Each point P of the plane has associated with it a pair of numbers called **coordinates**. The coordinates are defined in terms of the perpendicular distances from the axes to the point.

DEFINITION 3 *The x coordinate, or **abscissa**, of a point P is the directed distance from the y axis to the point. The y coordinate, or **ordinate**, of a point P is the directed distance from the x axis to the point.*

A point whose abscissa is x and whose ordinate is y is designated by (x,y), in that order, the abscissa always coming first. Hence the coordinates of a point are an ordered pair of numbers. Although a pair of coordinates determines a point, the coordinates themselves are often referred to as a point.

We assume that to any pair of real numbers (coordinates) there corresponds one definite point. Conversely, we assume that to each point of the plane there corresponds one definite pair of coordinates. This relation of points on a plane and pairs of real numbers is called a one-to-one correspondence. The device which we have described for obtaining this correspondence is called a **rectangular coordinate system**.

A point of given coordinates is **plotted** by measuring the proper distances from the axes and marking the point thus located. For example, if the coordinates of a point are $(-4,3)$, the abscissa -4 means the point is 4 units to the left of the y axis and the ordinate 3 (plus sign understood) means the point is 3 units above the x axis. Consequently, we locate the point by going from the origin 4 units to the left along the x axis and then 3 units upward parallel to the y axis (Fig. 5).

The coordinate axes divide their plane into four parts, called **quadrants**, which are numbered I to IV in Fig. 4. The coordinates of a point in the first quadrant are both positive, which is indicated in the figure by $(+, +)$. The signs of the coordinates in each of the other quadrants are similarly indicated.

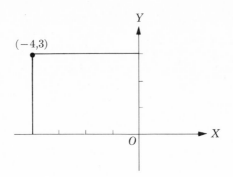

Figure 5

4. DISTANCE BETWEEN TWO POINTS

In many problems the distance between two points of the coordinate plane is required. The distance between any two points, or the length of the line segment connecting them, can be determined from the coordinates of the points. We shall classify a line segment (or line) as **horizontal, vertical**, or **slant**, depending on whether the segment is parallel to the x axis, to the y axis, or to neither axis. In deriving appropriate formulas for the lengths of these kinds of segments, we shall use the idea of directed segments.

Let $P_1(x_1,y)$ and $P_2(x_2,y)$ be two points on a horizontal line, and let A be the point where the line cuts the y axis (Fig. 6). We have, by Theorem 1,

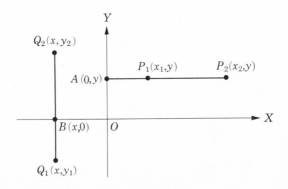

Figure 6

$$\overrightarrow{AP_1} + \overrightarrow{P_1P_2} = \overrightarrow{AP_2}$$

$$\overrightarrow{P_1P_2} = \overrightarrow{AP_2} - \overrightarrow{AP_1}$$

$$= x_2 - x_1$$

Similarly, for the vertical distance $\overrightarrow{Q_1Q_2}$, we have

$$\overrightarrow{Q_1Q_2} = \overrightarrow{Q_1B} + \overrightarrow{BQ_2}$$

$$= \overrightarrow{BQ_2} - \overrightarrow{BQ_1}$$

$$= y_2 - y_1$$

Hence the directed distance from a first point to a second point on a horizontal line is equal to the abscissa of the second point minus the abscissa of the first point. The distance is positive or negative according as the second point is to the right or left of the first point. A corresponding statement can be made relative to a vertical segment.

Inasmuch as the lengths of segments, without regard to direction, are often desired, we state a rule which gives results in positive quantities.

Rule 1. *The length of a horizontal line segment joining two points is the abscissa of the point on the right minus the abscissa of the point on the left.*

The length of a vertical line segment joining two points is the ordinate of the upper point minus the ordinate of the lower point.

We apply these rules to find the lengths of the line segments in Fig. 7:

$$|AB| = 5 - 1 = 4 \qquad\qquad |CD| = 6 - (-2) = 6 + 2 = 8$$

$$|EF| = 1 - (-4) = 1 + 4 = 5 \qquad |GH| = -2 - (-5) = -2 + 5 = 3$$

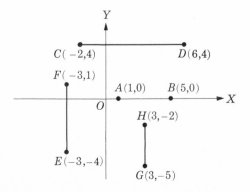

Figure 7

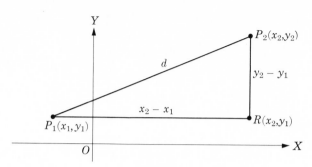

Figure 8

We next consider the points $P_1(x_1, y_1)$ and $P_2(x_2, y_2)$, which determine a slant line. Draw a line through P_1 parallel to the x axis and a line through P_2 parallel to the y axis (Fig. 8). These two lines intersect at the point R, whose abscissa is x_2 and whose ordinate is y_1. Hence

$$\overrightarrow{P_1 R} = x_2 - x_1 \qquad \text{and} \qquad \overrightarrow{R P_2} = y_2 - y_1$$

By the Pythagorean theorem,*

$$|P_1 P_2|^2 = (x_2 - x_1)^2 + (y_2 - y_1)^2$$

Denoting the length of the segment $P_1 P_2$ by d, we have the formula

$$d = \sqrt{(x_2 - x_1)^2 + (y_2 - y_1)^2}$$

The positive square root is chosen because we shall be interested only in the magnitude of the segment. We state this distance formula in words.

Rule 2. *To find the distance between two points, add the square of the difference of the abscissas to the square of the difference of the ordinates and take the positive square root of the sum.*

In employing the distance formula, either point may be designated by (x_1, y_1) and the other by (x_2, y_2). This results from the fact that the two differences involved are squared. The square of the difference of the two numbers is unchanged when the order of subtraction is reversed.

Example. Find the lengths of the sides of the triangle (Fig. 9) with the vertices $A(-2, -3)$, $B(6, 1)$, and $C(-2, 5)$.

---

* The **Pythagorean theorem** states that the *sum of the squares on the perpendicular sides of a right triangle is equal to the square on the hypotenuse.* That is, if a and b are the lengths of the perpendicular sides and c is the length of the hypotenuse, then $a^2 + b^2 = c^2$.

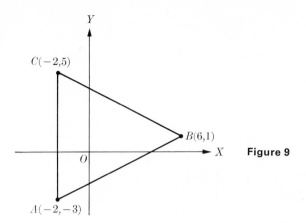

Figure 9

Solution. The abscissas of A and C are the same, and therefore side AC is vertical. The length of the vertical side is the difference of the ordinates. The other sides are slant segments, and the general distance formula yields their lengths. Hence we get

$$|AC| = 5 - (-3) = 5 + 3 = 8$$

$$|AB| = \sqrt{(6+2)^2 + (1+3)^2} = \sqrt{80} = 4\sqrt{5}$$

$$|BC| = \sqrt{(6+2)^2 + (1-5)^2} = \sqrt{80} = 4\sqrt{5}$$

The lengths of the sides show that the triangle is isosceles.

EXERCISES

1. Plot the points $A(-2,0)$, $B(1,0)$, and $C(5,0)$. Then find the directed distances $\overrightarrow{AB}$, $\overrightarrow{AC}$, $\overrightarrow{BC}$, $\overrightarrow{BA}$, $\overrightarrow{CA}$, and $\overrightarrow{CB}$.

2. Given the points $A(2,-3)$, $B(2,0)$, and $C(2,4)$, find the directed distances $\overrightarrow{AB}$, $\overrightarrow{BA}$, $\overrightarrow{AC}$, $\overrightarrow{CA}$, $\overrightarrow{BC}$, and $\overrightarrow{CB}$.

3. Plot the points $A(-2,\ 0)$, $B(2,0)$, and $C(5,0)$, and verify the following equations by numerical substitutions:

$$\overrightarrow{AB} + \overrightarrow{BC} = \overrightarrow{AC} \qquad \overrightarrow{AC} + \overrightarrow{CB} = \overrightarrow{AB} \qquad \overrightarrow{BA} + \overrightarrow{AC} = \overrightarrow{BC}$$

Find the distances between the pairs of points in Exercises 4 through 9.

4. $(1,3)$, $(4,7)$ 5. $(-3,4)$, $(2,-8)$

6. $(-2,-3)$, $(1,0)$ 7. $(5,-12)$, $(0,0)$

8. $(0,-4)$, $(3,0)$ 9. $(2,7)$, $(-1,4)$

In each of Exercises 10 through 13 draw the triangle with the given vertices and find the lengths of the sides.

10. $A(1,-1)$, $B(4,-1)$, $C(4,3)$ 11. $A(-1,2)$, $B(2,4)$, $C(0,5)$

12. $A(0,0)$, $B(-2,5)$, $C(2,-3)$ 13. $A(-3,0)$, $B(0,3)$, $C(-4,0)$

Draw the triangles in Exercises 14 through 17 and show that each is isosceles.

14. $A(6,1)$, $B(2,-4)$, $C(-2,1)$ 15. $A(6,4)$, $B(3,0)$, $C(-1,3)$

16. $A(1,-1)$, $B(1,7)$, $C(8,3)$ 17. $A(-3,-3)$, $B(3,3)$, $C(-4,4)$

Show that the triangles in Exercises 18 through 21 are right triangles.

18. $A(1,4)$, $B(10,6)$, $C(2,2)$ 19. $A(-2,1)$, $B(5,-2)$, $C(3,3)$

20. $A(0,4)$, $B(-3,-3)$, $C(2,-1)$ 21. A $(4,-3)$, $B(0,0)$, $C(3,4)$

22. Show that $A(-1,0)$, $B(3,0)$, and $C(1,2\sqrt{3})$ are vertices of an equilateral triangle.

23. Show that $A(-\sqrt{3},2)$, $B(2\sqrt{3},-1)$, and $C(2\sqrt{3},5)$ are vertices of an equilateral triangle.

24. Given the points $A(1,0)$, $B(5,3)$, $C(2,7)$, and $D(-2,4)$, show that the sides of the quadrilateral $ABCD$ are equal.

25. Determine whether the points $(-5,7)$, $(2,6)$, and $(1,-1)$ are all the same distance from $(-2,3)$.

26. Prove that the points $(-2,6)$, $(5,3)$, $(-1,-11)$, and $(-8,-8)$ are vertices of a rectangle.

Determine, by the distance formula, whether the points in each of Exercises 27 through 30 lie on a straight line.

27. $(3,2)$, $(0,0)$, $(9,6)$ 28. $(2,1)$, $(-1,2)$, $(5,0)$

29. $(-4,0)$, $(0,2)$, $(9,7)$ 30. $(-1,-1)$, $(6,-4)$, $(-10,3)$

31. If the point $(x,3)$ is equidistant from $(3,-2)$ and $(7,4)$, find x.

32. Find the point on the y axis which is equidistant from $(-5,-2)$ and $(3,2)$.

33. Plot a point $P_1(x_1,y_1)$ in the first quadrant and a point $P_2(x_2,y_2)$ in the fourth quadrant. Draw a suitable right triangle with P_1P_2 as the hypotenuse. From the diagram **derive** the distance formula.

34. Let the vertices of a triangle, reading counterclockwise, be $A(x_1,y_1)$, $B(x_2,y_2)$, and $C(x_3,y_3)$, as in Fig. 10. Observe that $DECA$, $EFBC$, and $DFBA$ are trapezoids. The sum of the areas of the first two trapezoids minus the area of the third trapezoid is equal to the area of triangle ABC. Recalling that the area of a trapezoid is equal to half the sum of the parallel sides times the altitude, we have area $DECA = \frac{1}{2}(y_1 + y_3)(x_3 - x_1)$. With this start, show that the area S of the triangle is

$$S = \tfrac{1}{2}[x_1(y_2 - y_3) - y_1(x_2 - x_3) + (x_2 y_3 - x_3 y_2)]$$

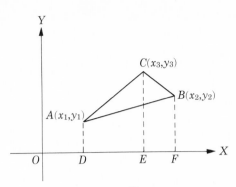

Figure 10

By expanding the determinant, show that the area of the triangle can be expressed as

$$S = \tfrac{1}{2} \begin{vmatrix} x_1 & y_1 & 1 \\ x_2 & y_2 & 1 \\ x_3 & y_3 & 1 \end{vmatrix}$$

What change is made in this result if the vertices are numbered in the clockwise direction?

Using the determinant of Exercise 34, find the area of each triangle having the given vertices.

35. $A(0,0)$, $B(6,0)$, $C(4,3)$

36. $A(-2,4)$, $B(2,-6)$, $C(5,4)$

37. $A(2,7)$, $B(1,1)$, $C(10,8)$

38. $A(5,-1)$, $B(-1,4)$, $C(3,6)$

39. $A(1,5)$, $B(6,1)$, $C(8,7)$

40. $A(-2,-3)$, $B(3,2)$, $C(-1,-8)$

5. INCLINATION AND SLOPE OF A LINE

Let a line intersect the x axis at the point M, and consider the angle which has one side to the right of M along the x axis and the other side upward along the line. We call these sides, respectively, the initial side and the terminal side. The **inclination** of the line is defined as the angle, measured counterclockwise, from the initial side to the terminal side. The inclination of a horizontal line is defined to be zero. In accordance with these definitions, the inclination θ of a line is such that $0° \leqslant \theta < 180°$.

In Fig. 11 the angle θ is the inclination of the line, MX is the initial side, and ML is the terminal side.

The **slope** of a line is defined as the tangent of its inclination. A line which leans to the right has a positive slope because the inclination is an acute angle. The slope of a line which leans to the left is negative. The slope of a horizontal line is zero. Vertical lines do not have a slope, however, since 90° has no tangent.

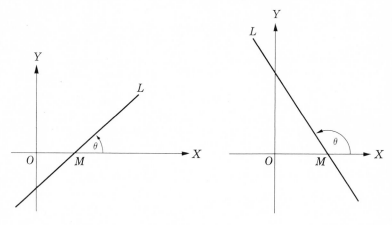

Figure 11

If the inclination of a nonvertical line is known, the slope can be determined by the use of a table of trigonometric functions. Conversely, if the slope of a line is known, its inclination can be found. In most problems, however, it is more convenient to deal with the slope of a line rather than with its inclination.

Example 1. Draw a line through the point $P(-2,2)$ with slope $-\frac{2}{3}$.

Solution. We move 3 units to the left of P and then 2 units upward. The line through the point thus located and the given point P clearly has the required slope (Fig. 12).

The definitions of inclination and slope lead immediately to a theorem concerning parallel lines. If two lines have the same slope, their inclinations are equal. Hence the lines are parallel. Conversely, if two nonvertical lines are parallel, they have equal inclinations, and consequently have equal slopes.

THEOREM 2   *Two nonvertical lines are parallel if, and only if, their slopes are equal.*

If the coordinates of two points on a line are known, we may find the slope of the line from the given coordinates. We now derive a formula for this purpose.

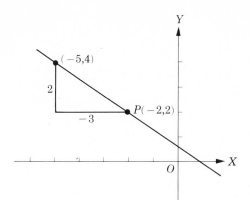

Figure 12

Let $P_1(x_1, y_1)$ and $P_2(x_2, y_2)$ be the two given points, and indicate the slope by m. Then, referring to Fig. 13, we have

$$m = \tan \theta = \frac{\overrightarrow{RP_2}}{\overrightarrow{P_1 R}} = \frac{y_2 - y_1}{x_2 - x_1}$$

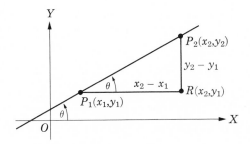

Figure 13

In Fig. 14 the line slants to the left. The quantities $y_1 - y_2$ and $x_2 - x_1$ are both positive and the angles θ and ϕ are supplementary. Consequently

$$\frac{y_1 - y_2}{x_2 - x_1} = \tan \phi = -\tan \theta$$

Therefore

$$m = \tan \theta = -\frac{y_1 - y_2}{x_2 - x_1} = \frac{y_2 - y_1}{x_2 - x_1}$$

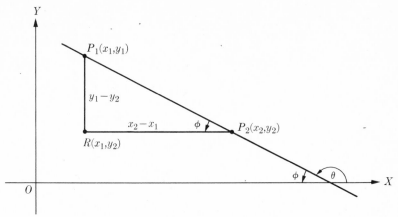

Figure 14

Hence the slope is determined in the same way for lines slanting either to the left or right.

THEOREM 3   *The slope m of a line passing through two given points $P_1(x_1, y_1)$ and $P_2(x_2, y_2)$ is equal to the difference of the ordinates divided by the difference of the abscissas taken in the same order; that is,*

$$m = \frac{y_2 - y_1}{x_2 - x_1}$$

This formula yields the slope if the two points are on a slant or a horizontal line. If the line is vertical, the denominator of the formula becomes zero, a result in keeping with the fact that slope is not defined for a vertical line. We observe further that either of the points may be designated as $P_1(x_1, y_1)$ and the other as $P_2(x_2, y_2)$, since

$$\frac{y_2 - y_1}{x_2 - x_1} = \frac{y_1 - y_2}{x_1 - x_2}$$

Example 2. Given the points $A(-1,-1)$, $B(5,0)$, $C(4,3)$, and $D(-2,2)$, show that $ABCD$ is a parallelogram.

Solution. We determine from the slopes of the sides if the figure is a parallelogram.

$$\text{Slope of } AB = \frac{0 - (-1)}{5 - (-1)} = \frac{1}{6} \qquad \text{Slope of } BC = \frac{3 - 0}{4 - 5} = -3$$

$$\text{Slope of } CD = \frac{2-3}{-2-4} = \frac{1}{6} \quad \text{Slope of } DA = \frac{2-(-1)}{-2-(-1)} = -3$$

The opposite sides have equal slopes, and therefore $ABCD$ is a parallelogram.

6. ANGLE BETWEEN TWO LINES

Two intersecting lines form two pairs of equal angles, and an angle of one pair is the supplement of an angle of the other pair. We shall show how to find a measure of each angle in terms of the slopes of the lines. Noticing Fig. 15 and

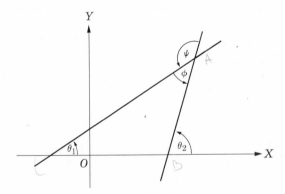

Figure 15

recalling that an exterior angle of a triangle is equal to the sum of the remote interior angles, we see that

$$\phi + \theta_1 = \theta_2 \quad \text{or} \quad \phi = \theta_2 - \theta_1$$

Using the formula for the tangent of the difference of two angles, we find

$$\tan \phi = \tan (\theta_2 - \theta_1) = \frac{\tan \theta_2 - \tan \theta_1}{1 + \tan \theta_1 \tan \theta_2}$$

If we let $m_2 = \tan \theta_2$ and $m_1 = \tan \theta_1$, then we have

$$\tan \phi = \frac{m_2 - m_1}{1 + m_1 m_2}$$

where m_2 is the slope of the terminal side and m_1 is the slope of the initial side, and ϕ is measured in a counterclockwise direction.

The angle ψ is the supplement of ϕ, and therefore

$$\tan \psi = -\tan \phi = \frac{m_1 - m_2}{1 + m_1 m_2}$$

This formula for $\tan \psi$ is the same as the one for $\tan \phi$ except that the terms in the numerator are reversed. We observe from the diagram, however, that the terminal side of ψ is the initial side of ϕ and the initial side of ψ is the terminal side of ϕ, as indicated by the counterclockwise arrows. Hence the numerator for $\tan \psi$ is equal to the slope of the terminal side of ψ minus the slope of the initial side of ψ. The same wording holds for $\tan \phi$; that is, the numerator for $\tan \phi$ is equal to the slope of the terminal side of ϕ minus the slope of the initial side of ϕ. Accordingly, in terms of the slopes of the initial and terminal sides, the tangent of either angle may be found by the same rule. We state this conclusion as a theorem.

THEOREM 4   *If ϕ is an angle, measured counterclockwise, between two lines, then*

$$\tan \phi = \frac{m_2 - m_1}{1 + m_1 m_2} \tag{1}$$

where m_2 is the slope of the terminal side and m_1 is the slope of the initial side.

This formula will not apply if either of the lines is vertical, since a vertical line does not possess slope. For this case the problem would be that of finding the angle, or a trigonometric function of the angle, which a line of known slope makes with the vertical. Hence no new formula is needed.

For any two slant lines which are not perpendicular, Eq. (1) will yield a definite number as the value of $\tan \phi$. Conversely, if the slopes of the lines are such that the formula yields a definite value, the lines could not be perpendicular, because the tangent of a right angle does not exist. Since the formula fails to yield a value only when the denominator is equal to zero, it appears that the lines are perpendicular when and only when $1 + m_1 m_2 = 0$ or

$$m_2 = -\frac{1}{m_1}$$

We note, additionally, that if α_2 and α_1 are the inclinations of slant lines which are perpendicular, then

$$\alpha_2 = \alpha_1 + 90° \qquad \text{or} \qquad \alpha_2 = \alpha_1 - 90°$$

In either case, $\tan \alpha_2 = -\cot \alpha_1$ and $m_2 = -\dfrac{1}{m_1}$.

THEOREM 5   *Two slant lines are perpendicular if, and only if, the slope of one is the negative reciprocal of the slope of the other.*

Perpendicularity of two lines occurs, of course, if one line is parallel to the x axis and the other is parallel to the y axis. The slope of the line parallel to the x axis is zero, but the line parallel to the y axis does not possess slope.

Example. Find the tangents of the angles of the triangle whose vertices are $A(3,-2)$, $B(-5,8)$, and $C(4,5)$. Then refer to a table and express each angle to the nearest degree.

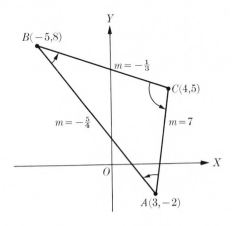

Figure 16

Solution. We first find the slope of each side. Thus, from Fig. 16,

$$\text{Slope of } AB = \frac{-2-8}{3-(-5)} = \frac{-5}{4}$$

$$\text{Slope of } BC = \frac{8-5}{-5-4} = -\frac{1}{3}$$

$$\text{Slope of } AC = \frac{-2-5}{3-4} = 7$$

We now substitute in Eq. (1) and get

$$\tan A = \frac{-\frac{5}{4}-7}{1+(-\frac{5}{4})(7)} = \frac{33}{31} = 1.06 \qquad A = 47°$$

$$\tan B = \frac{-\frac{1}{3}-(-\frac{5}{4})}{1+(-\frac{1}{3})(-\frac{5}{4})} = \frac{11}{17} = 0.647 \quad B = 33°$$

$$\tan C = \frac{7-(-\frac{1}{3})}{1+7(-\frac{1}{3})} = \frac{22}{-4} = -5.5 \qquad C = 100°$$

EXERCISES

Draw a line segment through the given point with the slope indicated in Exercises 1 through 4.

1. $(2,-3)$, $m = \frac{3}{4}$

2. $(4,0)$, $m = -3$

3. $(-1,-4)$, $m = -\frac{5}{3}$

4. $(3,5)$, $m = -5$

5. Give the slopes for the inclinations (a) $45°$, (b) $0°$, (c) $60°$, (d) $120°$, (e) $135°$. Find the slope of the line passing through the two points in each of Exercises 6 through 11. Find also the angle of inclination to the nearest degree. (See Table II in the Appendix.)

6. $(3,2)$, $(7,3)$

7. $(6,-13)$, $(0,3)$

8. $(4,-8)$, $(-7,3)$

9. $(4,5)$, $(-2,-3)$

10. $(0,-9)$, $(20,3)$

11. $(4,11)$, $(-8,-3)$

12. Show that each of the following sets of four points are vertices of the parallelogram $ABCD$.

   a) $A(2,0)$, $B(6,0)$, $C(4,3)$, $D(0,3)$
   b) $A(-2,2)$, $B(6,0)$, $C(5,-3)$, $D(-3,-1)$
   c) $A(0,-2)$, $B(4,-6)$, $C(12,-1)$, $D(8,3)$
   d) $A(-1,0)$, $B(5,2)$, $C(8,7)$, $D(2,5)$

13. Verify that each triangle with the given points as vertices is a right triangle by showing that the slope of one of the sides is the negative reciprocal of the slope of another side.

   a) $(3,-4)$, $(3,4)$, $(-1,0)$
   b) $(-1,1)$, $(3,-7)$, $(3,3)$
   c) $(8,1)$, $(1,-2)$, $(6,-4)$
   d) $(1,5)$, $(-6,7)$, $(-3,-9)$
   e) $(0,0)$, $(3,-2)$, $(2,3)$
   f) $(0,0)$, $(17,0)$, $(1,4)$

14. Show that the four points in each of the following sets are vertices of a rectangle.

   a) $(-5,3)$, $(-1,-2)$, $(4,2)$, $(0,7)$
   b) $(1,2)$, $(6,-3)$, $(9,0)$, $(4,5)$
   c) $(0,0)$, $(2,6)$, $(-1,7)$, $(-3,1)$
   d) $(5,-2)$, $(7,5)$, $(0,7)$, $(-2,0)$
   e) $(4,2)$, $(3,9)$, $(-4,8)$, $(-3,1)$
   f) $(4,6)$, $(0,0)$, $(3,-2)$, $(7,4)$

Using slopes, determine which of the sets of three points, in Exercises 15 through 18, lie on a straight line.

15. $(3,0)$, $(0,-2)$, $(9,4)$

16. $(2,1)$, $(-1,2)$, $(5,0)$

17. $(-4,-1)$, $(0,1)$, $(9,6)$

18. $(-1,-2)$, $(6,-5)$, $(-10,2)$

Find the tangents of the angles of the triangle ABC in Exercises 19 through 22.
Also find the angles to the nearest degree.

19. $A(-3,1)$, $B(3,5)$, $C(-1,6)$ 20. $A(-2,1)$, $B(1,3)$, $C(6,-7)$

21. $A(-3,1)$, $B(4,2)$, $C(2,3)$ 22. $A(1,3)$, $B(-2,-4)$, $C(3,-2)$

23. The line through the points $(4,3)$ and $(-6,0)$ intersects the line through $(0,0)$ and $(-1,5)$. Find the angles of intersection.

24. Two lines passing through $(2,3)$ make an angle of $45°$. If the slope of one of the lines is 2, find the slope of the other (two solutions).

25. What acute angle does a line of slope $-\frac{2}{3}$ make with a vertical line?

26. Find y if the slope of the line segment joining $(3,-2)$ to $(4,y)$ is -3.

27. The line segment drawn from $P(x,3)$ to $(4,1)$ is perpendicular to the segment drawn from $(-5,-6)$ to $(4,1)$. Find the value of x.

Find an equation in x and y if the point $P(x,y)$ is to satisfy the conditions in Exercises 28 through 31.

28. The line segment joining $P(x,y)$ and $(1,5)$ has slope 3.

29. The line segment joining $P(x,y)$ to $(2,4)$ is parallel to the segment joining $(-2,-1)$ and $(6,8)$.

30. The line segment joining $P(x,y)$ to $(2,4)$ is perpendicular to the segment joining $(-2,-1)$ and $(6,8)$.

31. The point $P(x,y)$ is on the line passing through $(2,-5)$ and $(7,1)$.

7. DIVISION OF A LINE SEGMENT

In this section we shall show how to find the coordinates of a point which divides a line segment into two parts which have a specified relation. We first find formulas for the coordinates of the point which is midway between two points of given coordinates.

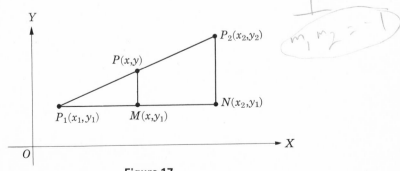

Figure 17

Let $P_1(x_1,y_1)$ and $P_2(x_2,y_2)$ be the extremities of a line segment, and let $P(x,y)$ be the midpoint of P_1P_2. From similar triangles (Fig. 17), we have

$$\frac{\overrightarrow{P_1P}}{\overrightarrow{P_1P_2}} = \frac{\overrightarrow{P_1M}}{\overrightarrow{P_1N}} = \frac{\overrightarrow{MP}}{\overrightarrow{NP_2}} = \frac{1}{2}$$

Hence

$$\frac{\overrightarrow{P_1M}}{\overrightarrow{P_1N}} = \frac{x-x_1}{x_2-x_1} = \frac{1}{2} \quad \text{and} \quad \frac{\overrightarrow{MP}}{\overrightarrow{NP_2}} = \frac{y-y_1}{y_2-y_1} = \frac{1}{2}$$

Solving for x and y gives

$$\boxed{x = \frac{x_1+x_2}{2} \qquad y = \frac{y_1+y_2}{2}}$$

THEOREM 6   *The abscissa of the midpoint of a line segment is half the sum of the abscissas of the endpoints; the ordinate is half the sum of the ordinates.*

This theorem may be generalized by letting $P(x,y)$ be any division point of the line through P_1 and P_2. If the ratio of $\overrightarrow{P_1P}$ to $\overrightarrow{P_1P_2}$ is a number r instead of $\frac{1}{2}$, then

$$\frac{\overrightarrow{P_1P}}{\overrightarrow{P_1P_2}} = \frac{x-x_1}{x_2-x_1} = r \quad \text{and} \quad \frac{\overrightarrow{P_1P}}{\overrightarrow{P_1P_2}} = \frac{y-y_1}{y_2-y_1} = r$$

These equations, when solved for x and y, give

$$\boxed{x = x_1 + r(x_2-x_1) \qquad y = y_1 + r(y_2-y_1)} \tag{2}$$

If P is between P_1 and P_2, as in Fig. 17, the directed distances $\overrightarrow{P_1P}$ and $\overrightarrow{P_1P_2}$ have the same sign, and their ratio r is positive and less than 1. If, however, P is a point on the segment P_1P_2 extended through P_2, then the length of the segment P_1P is greater than the length of P_1P_2, and r is greater than 1. Conversely, if r is greater than 1, the formulas yield the coordinates of a point on the extension of the segment through P_2. In order to find a point P on the segment extended in the other direction and still use a positive value for r, we designate the two given points P_1 and P_2 so that P_1 is farther from P than P_2 is.

Example 1. The points $P_1(-4,3)$ and $P_2(2,7)$ determine a line segment. Find (a) the coordinates of the midpoint of the segment, (b) the coordinates of the trisection point nearer P_2.

Solution. Applying the midpoint formulas, we have

$$x = \frac{x_1 + x_2}{2} = \frac{-4 + 2}{2} = -1 \qquad y = \frac{y_1 + y_2}{2} = \frac{3 + 7}{2} = 5$$

To obtain the trisection point nearer P_2, we use $r = \frac{2}{3}$ in Eq. (2). Thus

$$x = x_1 + r(x_2 - x_1) = -4 + \tfrac{2}{3}(2 + 4) = 0$$

$$y = y_1 + r(y_2 - y_1) = 3 + \tfrac{2}{3}(7 - 3) = \tfrac{17}{3}$$

Hence the coordinates of the midpoint are $(-1, 5)$ and of the desired trisection point $(0, \frac{17}{3})$.

Example 2. A point $P(x, y)$ is on the line passing through $A(-2, 5)$ and $B(4, 1)$. Find (a) the coordinates of P if it is twice as far from A as from B, (b) the coordinates of P if it is three times as far from B as from A.

Solution. a) Since $\overrightarrow{AP} = 2(\overrightarrow{BP})$, it follows that $\overrightarrow{BP} = \overrightarrow{AB}$ (Fig. 18).

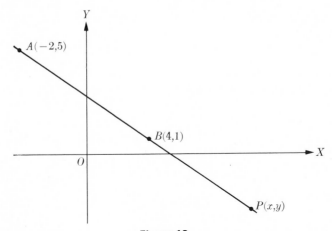

Figure 18

Hence the ratio of $\overrightarrow{AP}$ to $\overrightarrow{AB}$ is 2. Accordingly, we treat A as $P_1(x_1, y_1)$ of the formulas and B as $P_2(x_2, y_2)$. Thus, using $r = 2$, we write

$$x = -2 + 2(4 + 2) = 10 \qquad y = 5 + 2(1 - 5) = -3$$

Solution. b) For this position of P we have $\overrightarrow{BP} = 3(\overrightarrow{AP})$, and therefore $\overrightarrow{BA} = 2(\overrightarrow{AP})$ (Fig. 19). Hence r, the ratio of $\overrightarrow{BP}$ to $\overrightarrow{BA}$, is equal to $\frac{3}{2}$. Since P is farther from B than from A, we use B as P_1, and A and P_2 and obtain

$$x = 4 + \tfrac{3}{2}(-2 - 4) = -5 \qquad y = 1 + \tfrac{3}{2}(5 - 1) = 7$$

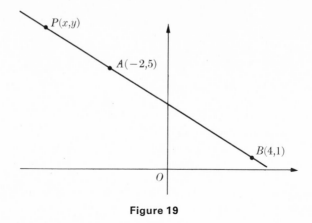

Figure 19

8. ANALYTIC PROOFS OF GEOMETRIC THEOREMS

By the use of a coordinate system, many theorems of geometry can be proved with surprising simplicity and directness. We illustrate the procedure in the following examples.

Example 1. Prove that the diagonals of a parallelogram bisect each other.

Proof. We first draw a parallelogram and then introduce a coordinate system. A judicious location of the axes relative to the figure makes the writing of the coordinates of the vertices easier and also simplifies the algebraic operations involved in making the proof. Therefore we choose a vertex as the origin and a coordinate axis along a side of the parallelogram (Fig. 20). Then we write the

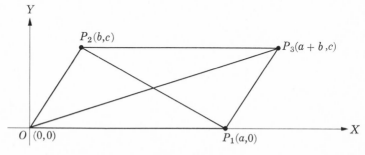

Figure 20

coordinates of the vertices as $O(0,0)$, $P_1(a,0)$, $P_2(b,c)$, and $P_3(a + b,c)$. It is essential that the coordinates of P_2 and P_3 express the fact that P_2P_3 and OP_1 are parallel and have the same length. This is achieved by making the ordinates of P_2 and P_3 the same and making the abscissa of P_3 exceed the abscissa of P_2 by a.

To show that OP_3 and P_1P_3 bisect each other, we find the coordinates of the midpoint of each diagonal.

$$\text{Midpoint of } OP_3: \quad x = \frac{a+b}{2} \qquad y = \frac{c}{2}$$

$$\text{Midpoint of } P_1P_2: \quad x = \frac{a+b}{2} \qquad y = \frac{c}{2}$$

Since the midpoint of each diagonal is $\left(\dfrac{a+b}{2}, \dfrac{c}{2}\right)$ the theorem is proved.

Note. In making a proof it is essential that a general figure be used. For example, neither a rectangle nor a rhombus (a parallelogram with all sides equal) should be used for a parallelogram. A proof of a theorem based on a special case would not constitute a general proof.

Example 2. The points $A(2,-4)$, $B(8,4)$, and $C(0,6)$ are vertices of a triangle. The line segment joining a vertex of a triangle and the midpoint of the opposite side is called a **median**. Find the coordinates of the point on each median which is two-thirds of the way from the vertex to the midpoint of the opposite side.

Solution. Figure 21 shows the triangle and the coordinates of the midpoints

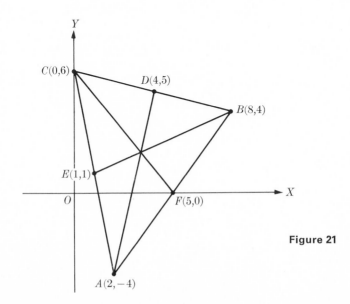

Figure 21

of the sides. Then using $r = \frac{2}{3}$ in the division formula, Eq. (2), we get for the medians AD, BE, and CF, respectively,

$$x = 2 + \tfrac{2}{3}(4 - 2) = \tfrac{10}{3} \qquad y = -4 + \tfrac{2}{3}(5 + 4) = 2$$

$$x = 8 + \tfrac{2}{3}(1 - 8) = \tfrac{10}{3} \qquad y = 4 + \tfrac{2}{3}(1 - 4) = 2$$

$$x = 0 + \tfrac{2}{3}(5 - 0) = \tfrac{10}{3} \qquad y = 6 + \tfrac{2}{3}(0 - 6) = 2$$

These results tell us that the medians are concurrent at the point $(\frac{10}{3}, 2)$.

EXERCISES

1. Find the coordinates of the midpoint of the line segment joining each of the following pairs of points:

   a) $(-3, 4)$, $(3, -4)$ b) $(2, 3)$, $(-6, -5)$
   c) $(3, -2)$, $(5, 4)$ d) $(9, 1)$, $(6, 4)$
   e) $(-4, 7)$, $(-6, -3)$ f) $(-8, -5)$, $(-3, -7)$

2. Find the coordinates of the midpoints of the sides of each of the triangles whose vertices are

   a) $(-1, -2)$, $(5, 2)$, $(1, 6)$ b) $(1, 1)$, $(3, 4)$, $(0, 6)$
   c) $(-3, 2)$, $(1, 8)$, $(-4, 7)$ d) $(0, -6)$, $(-2, -1)$, $(-5, 7)$

3. If the line segment connecting $(x_1, 7)$ and $(10, y_2)$ is bisected by the point $(8, 2)$, find the values of x_1 and y_2.

4. Find the lengths of the medians of the triangle whose vertices are $A(1, 5)$, $B(5, -1)$, $C(11, 9)$.

5. Find the coordinates of the midpoint of the hypotenuse of the right triangle whose vertices are $(1, 1)$, $(5, 2)$, and $(4, 6)$, and show that it is equidistant from each of the vertices.

6. Find the coordinates of the point which is $\frac{2}{3}$ of the distance from the first to the second point in each of the following:

   a) $(-1, -2)$, $(5, 4)$ b) $(4, -1)$, $(-1, 5)$
   c) $(0, 0)$, $(18, 5)$ d) $(1, 6)$, $(8, 12)$

7. Find the coordinates of the point which is $\frac{2}{5}$ of the way from $(3, 2)$ to $(-3, 5)$.

8. Find the coordinates of the point which divides the line segment connecting $(-1, 4)$ and $(2, -3)$ into two parts which have the ratio $\frac{3}{2}$.

9. Find the coordinates of P if it divides the line segment $A(1, -5)$, $B(7, -2)$ so that $\overrightarrow{AP} : \overrightarrow{PB} = 3 : 5$.

10. The line segment joining $A(1, 3)$ and $B(-2, -1)$ is extended through each

end by a distance equal to its original length. Find the coordinates of the new endpoints.

11. A line passes through $A(1,1)$ and $B(4,3)$. Find the coordinates of the point on the line which is (a) three times as far from A as from B, (b) three times as far from B as from A.

12. Find the coordinates of the intersection of the medians of the triangle whose vertices are $A(2,3)$, $B(-4,5)$, and $C(6,11)$.

Give analytic proofs of the following theorems.

13. The line segment connecting the midpoints of two sides of a triangle is parallel to the third side.

14. The line segment connecting the midpoints of two sides of a triangle is equal to one-half the third side.

15. The diagonals of a rectangle have the same length.

16. The midpoint of the hypotenuse of a right triangle is equidistant from the vertices.

17. The medians of a triangle intersect at a point two-thirds of the way from each vertex to the midpoint of the opposite side. Let the vertices be at the points $(0,0)$, $(a,0)$, and (b,c).

18. The line segment joining the midpoints of the nonparallel sides of a trapezoid is parallel to the bases and equal to half their sum.

19. The line segments connecting the midpoints of the opposite sides of a plane quadrilateral bisect each other.

20. The midpoints of the sides of a plane quadrilateral are the vertices of a parallelogram.

21. The diagonals of a square are perpendicular to each other.

22. The diagonals of an isosceles trapezoid are equal.

23. The lines drawn from a vertex of a parallelogram to the midpoints of the opposite sides trisect a diagonal.

9. RELATIONS AND FUNCTIONS

The concepts of relations and functions, which we shall meet in much of this book, are fundamental in many areas of mathematics. We introduce these ideas in this section starting with the definition of a relation.

DEFINITION 4   *A **relation** is a set of ordered pairs of numbers. The set of first elements (numbers) is called the **domain** of the relation and the set of second elements is called the **range** of the relation.*

Example 1. The set of number pairs

$$\{(-2,-2),\,(0,1),\,(2,2),\,(0,3),\,(4,0)\}$$

defines a relation. The set $\{-2,0,2,4\}$ is the domain and the set $\{-2,\,1,\,2,3,\,0\}$ is the range.

A relation may be determined by a given domain and an equation such that the second element of each number pair can be computed from the first. To illustrate, suppose that the second element of each pair (x,y) is given by the equation $y = 3x - 5$ and that the domain is the set of all real numbers. The relation defined by these conditions consists of infinitely many number pairs and therefore cannot be listed. We can, of course, find y for any particular value of x. Thus, the number pairs for $x = -1,0,2,4$ are

$$(-1,-8) \qquad (0,-5) \qquad (2,1) \qquad (4,7)$$

DEFINITION 5   *If for each number x belonging to a set of numbers X there is one and only one corresponding value y, then the set of ordered pairs (x,y) is called a **function**. The set X is called the **domain** of the function and the set of y values, which we indicate by Y, is called the **range**.*

We note that the letter x, in this definition, may be assigned any value from the set X, and y takes the corresponding value from the set Y. According to the following definition, x and y are variables.

DEFINITION 6   *A symbol, usually a letter, which may stand for any member of a specified set of objects is called a **variable**. If the set has only one member, the set is called a **constant**.*

The definition of a function requires that for each element of the domain there corresponds a unique element of the range. A relation, as illustrated in Example 1, may have more than one second element corresponding to a particular first element.

As a further illustration, consider the equations

$$y = 2\sqrt{x} \qquad \text{and} \qquad y = \pm 2\sqrt{x}$$

with the domain of each given by $X = \{1,2,3\}$. The first equation has a unique value y for each value of x, and therefore the equation, along with the given domain, defines a relation and also a function. The second equation has two y values corresponding to each x value. As a consequence, the resulting set of number pairs is a relation but not a function.

As we see, a relation or a function may be completely determined by an equation and a specific domain. Often, however, an equation is given with no accompanying domain. In all such cases, the domain is to consist of all real numbers x for which the equation yields real numbers y.

Example 2. Find the domain and range of the function defined by the equation $y = \sqrt{16 - x^2}$.

Solution. The variable x must not be assigned a value which makes the radicand negative. This is true because a square root of a negative number is not real. Because of this situation, the domain consists of the set of all real numbers from -4 to 4, inclusive. The resulting range is the set of all numbers from 0 to 4, inclusive.

10. GRAPHS OF RELATIONS AND FUNCTIONS

As previously hinted, the sciences of analytic geometry and algebra are correlated by means of coordinate systems. In plane analytic geometry, the underlying feature is the correspondence between an equation in x and y and a geometric figure. Accordingly, one problem is that of starting with an equation and finding the associated figure and, conversely, the other problem is that of passing from a given geometric figure to a corresponding equation. These two problems are sometimes called the **fundamental problems** of analytic geometry. We proceed now to deal with the first of the two problems. The figure corresponding to an equation is called the graph of the equation, which is defined as follows.

> DEFINITION 7   *The* **graph** *of a relation or function consists of the set of all points whose coordinates are ordered pairs of the relation or function. If the relation or function is defined by an equation, the graph of the equation is the same as the graph of the relation or function.*

Example 1. Draw the graph of the function defined by the equation $2x + 3y = 6$.

Solution. To draw the graph we assign values to x and find the corresponding values of y. The resulting ordered pairs are shown in the accompanying table. We plot each of the pairs as the abscissa and ordinate of a point. The points thus obtained appear to lie on a straight line (Fig. 22). The variables x and y are of the first degree in the given equation and therefore the equation is said to be **linear**. In the next chapter we shall prove that the graph of a linear equation in two variables is a straight line.

x	-3	-1.5	0	3	4.5	6
y	4	3	2	0	-1	-2

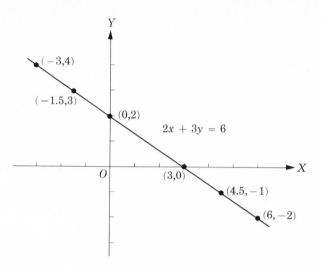

Figure 22

Example 2. Construct the graph of the equation

$$y = x^2 - 3x - 3$$

Solution. Any pair of numbers for x and y which satisfy the equation is called a **solution** of the equation. If a value is assigned to x, the corresponding value of y may be computed. Thus setting $x = -2$, we find $y = 7$. Several values of x and the corresponding values of y are shown in the table. These pairs of values, each constituting a solution, furnish a picture of the relation of x and y. A better representation is had, however, by plotting each value of x and the corresponding value of y as the abscissa and ordinate of a point and then drawing a smooth curve through the points thus obtained. This process is called **graphing the equation**, and the curve is called the **graph** of the equation.

x	-2	-1	0	1	1.5	2	3	4	5
y	7	1	-3	-5	-5.25	-5	-3	1	7

The plotted points (Fig. 23) extend from $x = -2$ to $x = 5$. Points corresponding to smaller and larger values of x could be plotted, and also any number of intermediate points could be located. But the plotted points show approximately where the intermediate points would be. Hence we can use a few points to draw a curve which is reasonably accurate. The curve shown here is called a **parabola**. We can, of course, draw only a part of the graph since the complete graph extends indefinitely far into the first and second quadrants.

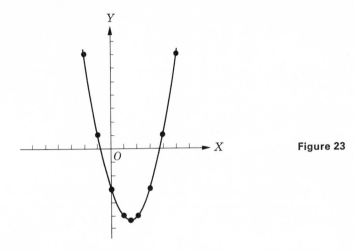

Figure 23

Example 3. Construct the graph of the relation defined by the equation

$$4x^2 + 9y^2 = 36$$

Solution. We solve the equation for y to obtain a suitable form for making a table of values. Thus, we get

$$y = \pm\tfrac{2}{3}\sqrt{9 - x^2}$$

We now see that x can take values only from -3 to 3; other values for x would yield imaginary values for y. The number pairs in the table yield points of the graph. The curve drawn through the points (Fig. 24) is called an **ellipse**.

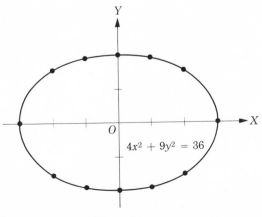

$$4x^2 + 9y^2 = 36$$

Figure 24

(Compare to the parabola shown in Fig. 23. Later we shall study curves of this kind in considerable detail.)

x	-3	-2	-1	0	1	2	3
y	0	±1.5	±1.9	±2	±1.9	±1.5	0

DEFINITION 8   *The abscissa of a point where a curve touches or crosses the x axis is called an **x intercept**, and the ordinate of a point where a curve touches or crosses the y axis is called a **y intercept**.*

To find the x intercepts, if any, of the graph of an equation, we set $y = 0$ and solve the resulting equation for x. Similarly, we set $x = 0$ and solve for y to find the y intercepts. Thus, the x intercepts of the equation

$$y + x^2 - 2x - 3 = 0$$

are -1 and 3, and the y intercept is 3. The intercepts will be helpful in drawing the graphs of the equations in the following exercise.

EXERCISES

Plot a few points and draw the graph of each equation. To find square roots of numbers see Table I in the Appendix.

1. $y = 4 - x$
2. $y = -2x$
3. $3x - 2y = 6$
4. $5x + 3y = 15$
5. $y = x^2$
6. $y^2 = x$
7. $y = x^2 - 2x - 1$
8. $y = x^2 + 4x$
9. $x = y^2 - 4$
10. $x^2 + y^2 = 9$
11. $9x^2 + 4y^2 = 36$
12. $x^2 + y^2 = 25$

11. THE EQUATION OF A GRAPH

Having obtained graphs of equations, we naturally surmise that a graph may have a corresponding equation. We shall consider the problem of writing the equation of a graph all of whose points are definitely fixed by given geometric conditions. This problem is the inverse of drawing the graph of an equation.

DEFINITION 9   *An equation in x and y which is satisfied by the coordinates of all points of a graph and only those points is said to be an equation of the graph.**

---

* A graph may be represented by more than one equation. For instance, the graph of $(x^2 + 1)(x + y) = 0$ and $x + y = 0$ is the same straight line. Sometimes, however, we shall speak of "the" equation when really we mean the simplest obtainable equation.

The procedure for finding the equation of a graph is straightforward. Each point $P(x,y)$ of the graph must satisfy the specified conditions. The desired equation can be written by requiring the point P to obey the conditions. The examples illustrate the method.

Example 1. A line with slope 2 passes through the point $(-3,4)$. Find the equation of the line.

Solution. **We** apply the formula for the slope of a line through two points (Section 5, Chapter 1). Thus, the slope through $P(x,y)$ and $(-3,4)$ is

$$m = \frac{y-4}{x-(-3)} = \frac{y-4}{x+3}$$

We equate this expression to the given slope. Hence

$$\frac{y-4}{x+3} = 2$$

or

$$2x - y + 10 = 0$$

Example 2. Find the equation of the set of all points equally distant from the y axis and $(4,0)$.

Solution. We take a point $P(x,y)$ of the graph (Fig. 25). Then, referring to the distance formula (Section 4, Chapter 1), we find the distance of P from the y axis to be the abscissa x, and the distance from the point $(4,0)$ to be

$$\sqrt{(x-4)^2 + y^2}$$

Equating the two distances, we obtain

$$\sqrt{(x-4)^2 + y^2} = x$$

By squaring both sides and simplifying, we get

$$y^2 - 8x + 16 = 0$$

Example 3. Find the equation of the set of all points which are twice as far from $(4,4)$ as from $(1,1)$.

Solution. We apply the distance formula to find the distance of a point $P(x,y)$ from each of the given points. Thus we obtain the expressions

$$\sqrt{(x-1)^2 + (y-1)^2} \qquad \text{and} \qquad \sqrt{(x-4)^2 + (y-4)^2}$$

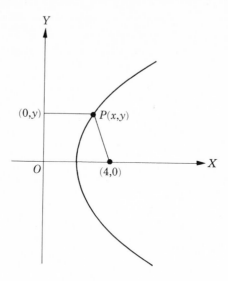

Figure 25

Since the second distance is twice the first, we have the equation

$$2\sqrt{(x-1)^2+(y-1)^2}=\sqrt{(x-4)^2+(y-4)^2}$$

Simplifying, we get

$$4(x^2-2x+1+y^2-2y+1)=x^2-8x+16+y^2-8y+16$$

or

$$x^2+y^2=8$$

The graph of the equation appears in Fig 26.

Example 4. Find the equation of the set of all points $P(x,y)$ such that the sum of the distances of P from $(-5,0)$ and $(5,0)$ is equal to 14.

Solution. Referring to Fig. 27, we get the equation

$$\sqrt{(x+5)^2+y^2}+\sqrt{(x-5)^2+y^2}=14$$

By transposing the second radical, squaring, and simplifying, we obtain the equation

$$7\sqrt{(x-5)^2+y^2}=49-5x$$

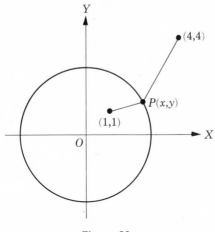

Figure 26

Squaring again and simplifying, we have the equation

$$24x^2 + 49y^2 = 1176$$

As shown in the figure, the x intercepts of the graph of this equation are $(-7,0)$ and $(7,0)$, and the y intercepts are $(0,-\sqrt{24})$ and $(0,\sqrt{24})$.

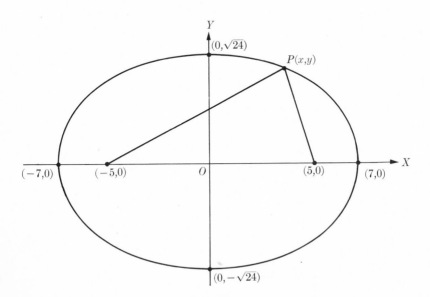

Figure 27

EXERCISES

Find the equations of the following lines.

1. The line passing through the origin with slope 2.
2. The line passing through $(2,4)$ with slope -3.
3. The line passing through $(-2,3)$ with slope $\frac{1}{2}$.
4. The line passing through $(-4,3)$ with slope $-\frac{2}{3}$.
5. The vertical line passing through $(6,0)$.
6. The horizontal line passing through $(-3,3)$.
7. The line 2 units below the x axis.
8. The line 3 units to the left of the y axis.

In Exercises 9 through 18 find the equation of the set of all points $P(x,y)$ which satisfy the given condition. From the equation sketch the graph.

9. $P(x,y)$ is equidistant from $(2,-4)$ and $(-1,5)$.
10. $P(x,y)$ is equidistant from $(0,0)$ and $(5,-3)$.
11. $P(x,y)$ is equidistant from the y axis and the point $(5,0)$.
12. $P(x,y)$ is equidistant from the point $(4,0)$ and the line $x = -4$.
13. $P(x,y)$ is twice as far from $(-4,4)$ as from $(-1,1)$.
14. $P(x,y)$ is twice as far from $(8,-8)$ as from $(2,-2)$.
15. $P(x,y)$ forms with $(-5,0)$ and $(5,0)$ the vertices of a right triangle with P the vertex of the right angle.
16. $P(x,y)$ forms with $(-3,0)$ and $(3,0)$ the vertices of a right triangle with P the vertex of the right angle.
17. The sum of the distances of $P(x,y)$ from $(-3,0)$ and $(3,0)$ is equal to 10.
18. The sum of the distances of $P(x,y)$ from $(0,-2)$ and $(0,2)$ is equal to 8.

Chapter 2
The Straight Line and the Circle

1. LINES AND FIRST-DEGREE EQUATIONS

The straight line is the simplest geometric curve. Despite its simplicity, the line is a vital concept of mathematics and enters into our daily experiences in numerous interesting and useful ways. In our study of the straight line, we shall first discover a correspondence between a line and a first-degree equation in x and y. The equation will represent the line, and the line will be the graph of the equation. With respect to this relationship, we state the following theorem.

> **THEOREM 1** *The equation of every straight line is expressible in terms of the first degree. Conversely, the graph of a first-degree equation is a straight line.*

Proof. Suppose we start with a fixed line in the coordinate plane. The line may be parallel to the y axis or not parallel to the y axis. First, let us consider a line L parallel to the y axis and at a distance a from the axis (Fig. 1). The abscissas of all points on the line are equal to a. So we see at once that an equation of the line is

$$\boxed{x = a} \tag{1}$$

Conversely, the equation $x = a$ is satisfied by the coordinates of every point on the line at a distance a from the y axis. Hence the line L is the graph of the equation.

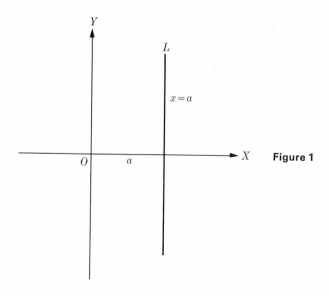

Figure 1

Next, we consider a line not parallel to the y axis (Fig. 2). Such a line has a slope, and the line intersects the y axis at a point whose abscissa is zero. We let m stand for the slope and b for the ordinate of the intersection point. Then if (x,y) are the coordinates of any other point of the line, we apply the formula for the slope of a line through two points and have the equation

$$\frac{y-b}{x-0} = m$$

which reduces to

$$\boxed{y = mx + b} \tag{2}$$

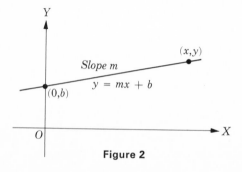

Figure 2

This equation makes evident the slope and y intercept of the line which it represents, and is said to be in the **slope-intercept form.**

We started with a fixed line and obtained Eq. (2). Now suppose we start with the equation and determine its graph. Clearly the equation is satisfied by $x = 0$ and $y = b$. Let (x,y) be a point other than $(0,b)$ on the graph. This means that x and y satisfy Eq. (2) and consequently the equivalent equation

$$\frac{y-b}{x-0} = m$$

This equation tells us that the point (x,y) of the graph must be on the line through $(0,b)$ with slope m. Hence any point whose coordinates satisfy Eq. (2) is on the line.

We now have shown that the equations of all lines can be expressed in the forms (1) and (2) and, conversely, that the graphs of such equations are straight lines. To complete the proof of the theorem, we need only to point out that the general first-degree equation

$$Ax + By + C = 0 \tag{3}$$

can be put in the form (1) or (2). We assume that A, B, and C are constants with A and B not both zero. Accordingly, if $B = 0$, then $A \neq 0$, and Eq. (2) reduces to

$$x = -\frac{C}{A} \tag{4}$$

Next, if $B \neq 0$, we can solve Eq. (3) for y. Thus, we get

$$y = -\frac{A}{B}x - \frac{C}{B} \tag{5}$$

Equation (4) is in the form of Eq. (1) with $a = -(C/B)$, and Eq. (5) is in the form of Eq. (2) with $m = -(A/B)$ and $b = -(C/B)$. This completes the proof of the theorem.

We note that Eq. (3), with $b \neq 0$, yields for each value of x a unique corresponding value for y. This means that the equation defines a function (Chapter 1, Definition 5) whose graph is a straight line not parallel to the y axis.

Example 1. Write the equation of the line with slope -3 and y intercept 4.

Solution. In the slope–intercept form (2), we substitute -3 for m and 4 for b. This gives at once the equation $y = -3x + 4$.

Example 2. Express the equation $4x - 3y - 11 = 0$ in the slope-intercept form.

Solution. The given equation is in the general form (3) and, when solved for y, reduces to an equation whose coefficient of x is the slope of the graph and whose constant term is the y intercept. So we solve for y and have readily the desired equation

$$y = \tfrac{4}{3}x - \tfrac{11}{3}$$

The slope–intercept form (2) represents a line which passes through the point $(0,b)$. The equation can be altered slightly to focus the attention on any other point of the line. If (x_1, y_1) is another point of the line, we can replace x by x_1 and y by y_1 in the form (2). This gives

$$y_1 = mx_1 + b \qquad \text{or} \qquad b = y_1 - mx_1$$

Substituting $y_1 - mx_1$ for b in Eq. (2), we have

$$y = mx + y_1 - mx_1$$

or, equivalently,

$$\boxed{y - y_1 = m(x - x_1)} \tag{6}$$

This equation tells us that the point (x_1, y_1) is on the line and that the slope is m. Accordingly, the equation is said to be in the **point–slope form.**

Example 3. Find the equation of the line which passes through $(2, -3)$ and has a slope of 5.

Solution. Using the point–slope form (6) with $x_1 = 2$, $y_1 = -3$, and $m = 5$, we get

$$y + 3 = 5(x - 2) \qquad \text{or} \qquad 5x - y = 13$$

Example 4. Find the equation of the line determined by the points $(-3, 4)$ and $(2, -2)$.

Solution. After finding the slope of the line through the given points (Section 5, Chapter 1), we apply the point-slope Eq. (6). Thus,

$$m = \frac{-2 - 4}{2 - (-3)} = -\frac{6}{5}$$

and

$$y - 4 = -\tfrac{6}{5}(x + 3) \qquad \text{or} \qquad 6x + 5y - 2 = 0$$

EXERCISES

By inspection, give the slope and intercepts of each line represented by the equations in Exercises 1 through 12. Reduce each equation to the slope-intercept form by solving for y.

1. $4x - y = 12$
2. $x - y = 7$
3. $x + y + 4 = 0$
4. $4x + 9y = 36$
5. $3x - 4y = 12$
6. $6x - 3y - 10 = 0$
7. $x + 7y = 11$
8. $2x + 3y = 14$
9. $7x + 3y + 6 = 0$
10. $2x - 8y = 5$
11. $8x + 3y = 4$
12. $3x + 3y = 1$

In Exercises 13 through 20, write the equation of the line determined by the slope m and y intercept b.

13. $m = 3, b = -4$
14. $m = 2, b = 3$
15. $m = -4, b = 5$
16. $m = -5, b = 0$
17. $m = \frac{2}{3}, b = -2$
18. $m = \frac{3}{2}, b = -6$
19. $m = 0, b = 7$
20. $m = 5, b = 0$

In Exercises 21 through 30, write the equation of the line which passes through the point A with the slope m. Draw the lines.

21. $A(3, 1), m = 2$
22. $A(-3, -5), m = 1$
23. $A(-2, 0), m = \frac{2}{3}$
24. $A(0, -3), m = \frac{3}{2}$
25. $A(-3, -6), m = -\frac{1}{2}$
26. $A(5, -2), m = -\frac{5}{2}$
27. $A(0, 3), m = 0$
28. $A(3, 0) \ m = 0$
29. $A(0, 0), m = -\frac{8}{3}$
30. $A(0, 0), m = \frac{2}{7}$

Find the equation of the line determined by the points A and B in Exercises 31 through 40. Check the answers by substitutions.

31. $A(3, -1), B(-4, 5)$
32. $A(1, 5), B(4, 1)$
33. $A(0, 2), B(4, -6)$
34. $A(-2, -4), B(3, 3)$
35. $A(3, -2), B(3, 7)$
36. $A(0, 0), B(3, -4)$
37. $A(5, -\frac{2}{3}), B(\frac{1}{2}, -2)$
38. $A(\frac{1}{2}, 5), B(\frac{3}{4}, -1)$
39. $A(-6, -1), B(4, -1)$
40. $A(0, 1), B(0, 0)$

2. OTHER FORMS OF FIRST-DEGREE EQUATIONS

The equation of a straight line may be expressed in several different forms such that each form exhibits certain geometric properties of the line. As we have seen, the slope–intercept form brings into focus the slope and y intercept and the point–slope form brings into focus the slope and a point of the line. In this section we shall derive two additional forms, each having its advantages in certain situations.

We first introduce an alternative form for the point–slope equation. To obtain this form, we substitute $-(A/B)$ for m in Eq. (6) and have

$$y - y_1 = -\frac{A}{B}(x - x_1)$$

Multiplying by B and transposing terms yields

$$\boxed{Ax + By = Ax_1 + By_1} \tag{7}$$

Example 1. Find the equation of the line of slope $\frac{2}{5}$ and passing through the point $(-3, 4)$.

Solution. Since the slope, $-(A/B)$, is equal to $\frac{2}{5}$, we let $A = 2$ and $B = -5$. These values along with $x_1 = -3$ and $y_1 = 4$ may be substituted in Eq. (7). Thus we have

$$2x - 5y = 2(-3) - 5(4) = -26$$

or, as a final result,

$$2x - 5y + 26 = 0$$

Notice that this equation is more directly obtainable by formula (7) than by the point–slope formula (6).

Suppose the x intercept of a line is a and the y intercept is b, where $a \neq 0$ and $b \neq 0$ (Fig. 3). Then the line passes through the point $(a, 0)$ and $(0, b)$, and consequently the slope is $-(b/a)$. Then, applying the point–slope formula, we find the equation

$$y - b = -\frac{b}{a}(x - 0)$$

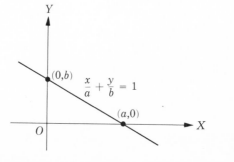

Figure 3

which can be reduced to

$$\frac{x}{a} + \frac{y}{b} = 1 \qquad (8)$$

This is called the **intercept form** because the intercepts are exhibited in the denominators.

Example 2. Write the equation of the line whose x intercept is 3 and whose y intercept is -5. We replace a by 3 and b by -5 in formula (8) and get at once

$$\frac{x}{3} + \frac{y}{-5} = 1 \qquad \text{or} \qquad 5x - 3y = 15$$

Example 3. Write the equation $4x - 9y = -36$ in the intercept form.

Solution. In order to have an equation with the left member equal to unity, as in formula (8), we divide the members of the given equation by -36. This gives the desired form

$$\frac{x}{-9} + \frac{y}{4} = 1$$

We could, of course, obtain this result by first finding the intercepts from the given equation and then substituting in formula (8).

Example 4. Find the equation of the line through the point $(2,-3)$ and perpendicular to the line defined by the equation $4x + 5y + 7 = 0$.

Solution. We let $5x - 4y$ be the left side of our equation. This will provide for the necessary slope of the perpendicular line. In order that the line shall pass through the given point, we apply formula (7) with $x_1 = 2$ and $y_1 = -3$. Thus we get

$$5x - 4y = 5(2) - 4(-3) \qquad \text{or} \qquad 5x - 4y = 22$$

Example 5. The ends of a line segment are at $C(7,-2)$ and $D(1,6)$. Find the equation of the perpendicular bisector of the segment CD.

Solution. The slope of CD is $-\frac{4}{3}$, and the midpoint is at $(4,2)$. The perpendicular bisector therefore has a slope of $\frac{3}{4}$ and passes through $(4,2)$. Employing formula (7) with $A = 3$, $B = -4$, $x_1 = 4$, and $y_1 = 2$, we obtain

$$3x - 4y = 3(4) - 4(2) \qquad \text{or} \qquad 3x - 4y = 4$$

EXERCISES

In Exercises 1 through 6, write the equation of the line through the point A with the slope m. Use formula (7).

1. $A(1,4)$, $m = \frac{2}{3}$
2. $A(-3,5)$, $m = -\frac{3}{4}$
3. $A(-4,0)$, $m = -\frac{5}{3}$
4. $A(7,1)$, $m = \frac{2}{5}$
5. $A(-6,-1)$, $m = -\frac{7}{9}$
6. $A(-2,-4)$, $m = \frac{6}{5}$

Write the equation of the line with x intercept a and y intercept b in Exercises 7 through 14.

7. $a = 2$, $b = 3$
8. $a = 1$, $b = 5$
9. $a = -3$, $b = 4$
10. $a = 7$, $b = -6$
11. $a = -4$, $b = -4$
12. $a = -\frac{2}{3}$, $b = 1$
13. $a = \frac{4}{5}$, $b = \frac{1}{2}$
14. $a = \frac{3}{4}$, $b = -\frac{4}{3}$

Express each equation, in Exercises 15 through 23, in the intercept form.

15. $x - 4y = 8$
16. $2x + 3y = 6$
17. $9y = 4x + 36$
18. $3x - 4y = 8$
19. $2x = 5y + 9$
20. $5x - 6 = 3y$
21. $\dfrac{2x}{3} - \dfrac{y}{2} = \dfrac{5}{6}$
22. $\dfrac{12x}{5} - \dfrac{3y}{5} = \dfrac{4}{3}$
23. $\dfrac{3x}{4} + \dfrac{5y}{2} = \dfrac{3}{2}$

24. Show that $Ax + By = D_1$ and $Bx - Ay = D_2$ are equations of perpendicular lines. Assume that A and B are not both zero.

In Exercises 25 through 32, find the equations of two lines through A, one parallel and the other perpendicular to the line defined by the given equation.

25. $A(4,1)$, $2x - 3y + 4 = 0$
26. $A(-1,2)$, $2x - y = 0$
27. $A(2,-3)$, $8x - y = 0$
28. $A(0,6)$, $2x - 2y = 1$
29. $A(7,0)$, $9x + y - 3 = 0$
30. $A(-4,0)$, $4x + 3y = 2$
31. $A(-1,1)$, $y = 1$
32. $A(3,5)$, $x = 0$

The points A, B, and C in Exercises 33 through 36 are vertices of triangles. For each triangle find:

   a) the equations of the medians and their intersection point,
   b) the equations of the altitudes and their intersection point,
   c) the equations of the perpendicular bisectors of the sides and the intersection point.

33. $A(0,0)$, $B(6,0)$, $C(4,4)$
34. $A(0,0)$, $B(9,2)$, $C(0,7)$
35. $A(1,0)$, $B(9,2)$, $C(3,6)$
36. $A(-2,3)$, $B(6,-6)$, $C(8,0)$

37. Show that the three intersection points involved in Exercise 33 are on a straight line. Do the same for the remaining exercises.

3. DIRECTED DISTANCE FROM A LINE TO A POINT

The directed distance from a line to a point can be found from the equation of the line and the coordinates of the point. For a vertical line and a horizontal line the distances are easily determined. We notice these special cases first, and then will consider a slant line. The vertical line L is at a distance a from the y axis and $P_1(x_1,y_1)$ is a point off the line (Fig. 4). The foot of the perpendicular

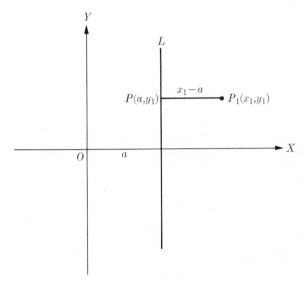

Figure 4

from P_1 to L is at the point $P(a,y_1)$. Since P_1 is to the right of L, we specify that the distance from P to P_1 is positive. Hence, we express this distance by the equation

$$\overrightarrow{PP_1} = x_1 - a \tag{9}$$

If P_1 were to the left of L, then $x_1 - a$ would be negative. Consequently Eq. (9) gives the directed distance from the line L if P_1 is any point to the right or the left of L. Similarly, the directed distance from a horizontal line $y = b$ to a point $P_1(x_1,y_1)$ is equal to $y_1 - b$.

We next consider a slant line L and a point $P_1(x_1,y_1)$, as shown in Fig. 5. We seek now the perpendicular distance from L to the point P_1. The line L' passes through P_1 and is parallel to L. The line through the origin perpendicular to L intersects L and L' at P and Q. We choose the general linear equation to

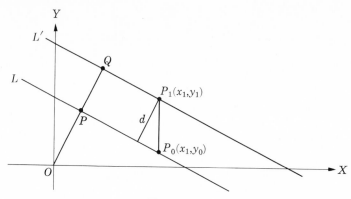

Figure 5

represent the line L. Then we may express the equations of L, L', and the perpendicular line, respectively, by

$$Ax + By + C = 0 \tag{10}$$

$$Ax + By + C' = 0 \tag{11}$$

$$Bx - Ay = 0 \tag{12}$$

The simultaneous solution equation of Eqs. (10) and (12) gives the coordinates of P, and the simultaneous solution of Eqs. (11) and (12) gives the coordinates of Q. These coordinates, as the student may verify, are

$$P\left(\frac{-AC}{A^2 + B^2}, \frac{-BC}{A^2 + B^2}\right) \quad \text{and} \quad Q\left(\frac{-AC'}{A^2 + B^2}, \frac{-BC'}{A^2 + B^2}\right)$$

We let d denote the perpendicular distance from the given line L to the point P_1. Then d is also equal to the distance from P to Q. To find the distance from P to Q, we use the distance formula of Section 4, chapter 1. Thus, we get

$$d^2 = |PQ|^2 = \frac{(C - C')^2 A^2}{(A^2 + B^2)^2} + \frac{(C - C')^2 B^2}{(A^2 + B^2)^2}$$

$$= \frac{(C - C')^2 (A^2 + B^2)}{(A^2 + B^2)^2} = \frac{(C - C')^2}{A^2 + B^2}$$

and

$$d = \frac{C - C'}{\pm\sqrt{A^2 + B^2}}$$

Since the line of Eq. (11) passes through $P_1(x_1, y_1)$, we have

$$Ax_1 + By_1 + C' = 0, \quad \text{and} \quad C' = -Ax_1 - By_1.$$

Hence, substituting for C', we get

$$d = \frac{Ax_1 + By_1 + C}{\pm\sqrt{A^2 + B^2}} \tag{13}$$

To remove the ambiguity of the sign in the denominator, let us agree that d is to be positive if P_1 is above the given line L and negative if P_1 is below the line. We can achieve this end by choosing the sign of the denominator the same as the sign of B. That is, the coefficient of y_1 will be $B/\sqrt{A^2 + B^2}$ when B is positive and will be $B/-\sqrt{A^2 + B^2}$ when B is negative. Therefore, in either case, the coefficient of y_1 will be positive. Referring again to the figure, we let $P_0(x_1, y_0)$ be a point on the line L such that $P_0 P_1$ is parallel to the y axis. Since P_0 is on the line L, we have the equation

$$\frac{Ax_1 + By_0 + C}{\pm\sqrt{A^2 + B^2}} = 0 \tag{14}$$

Now if we replace y_0 by y_1 in the left side of Eq. (14) we get a fraction which is not equal to zero. Letting P_1 be above the line L and letting D (a positive quantity) stand for the coefficient of y_0, we have the inequalities

$$y_1 > y_0 \quad \text{and} \quad Dy_1 > Dy_0$$

These inequalities tell us that the value of the new fraction is greater than zero. Similarly, if P_1 is below the line L, the value of the fraction becomes negative. We may therefore regard the distance from a slant line to a point as a directed distance.

On the basis of the preceding discussion, we state the following theorem:

THEOREM 2   *The directed distance from the slant line $Ax + By + C = 0$ to the point $P_1(x_1, y_1)$ is given by the formula*

$$\boxed{d = \frac{Ax_1 + By_1 + C}{\pm\sqrt{A^2 + B^2}}} \tag{15}$$

where the denominator is given the sign of B. The distance is positive if the point P_1 is above the line, and negative if P_1 is below the line.

Example 1. Find the distance from the line $5x = 12y + 26$ to the points $P_1(3, -5)$, $P_2(-4, 1)$, and $P_3(9, 0)$.

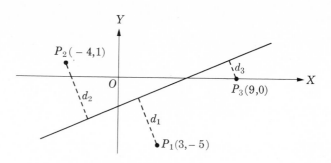

Figure 6

Solution. We write the equation in the form $5x - 12y - 26 = 0$, and apply formula (15) with the denominator negative. Then referring to Fig. (6), where the given line and the given points are plotted, we have

$$d_1 = \frac{5(3) - 12(-5) - 26}{-\sqrt{5^2 + 12^2}} = \frac{49}{-13} = -\frac{49}{13}$$

$$d_2 = \frac{5(-4) - 12(1) - 26}{-13} = \frac{-58}{-13} = \frac{58}{13}$$

$$d_3 = \frac{5(9) - 12(0) - 26}{-13} = \frac{19}{-13} = -\frac{19}{13}$$

Example 2. Find the distance between the parallel lines $15x + 8y + 68 = 0$ and $15x + 8y - 51 = 0$.

Solution. The distance between the lines can be found by computing the distance from each line to a particular point. To minimize the computations, we choose the origin as the particular point. Thus,

$$d_1 = \frac{15(0) + 8(0) + 68}{\sqrt{15^2 + 8^2}} = \frac{68}{17} = 4$$

$$d_2 = \frac{15(0) + 8(0) - 51}{17} = \frac{-51}{17} = -3$$

These distances reveal that the origin is 4 units above the first line and 3 units below the second line. Hence the given lines are 7 units apart.

An alternative method for this problem would be to find the distance from one of the lines to some point on the other line. The point $(3.4, 0)$ is on the second given line. So using this point and the first equation, we find

$$d = \frac{15(3.4) + 8(0) + 68}{17} = \frac{119}{17} = 7$$

Example 3. Find the equation of the bisector of the acute angle formed by the lines $x - 2y + 1 = 0$ and $x + 3y - 3 = 0$.

Solution. The undirected distances from the sides of the angle to a point on the bisector are the same. So we choose a point $P(x, y)$ on the bisector and denote the distances from the sides to P by d_1 and d_2 (Fig. 7). We observe from the

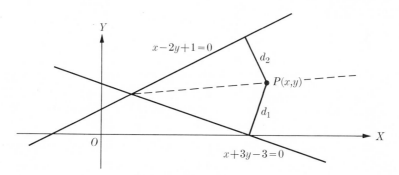

Figure 7

figure that P is above one of the sides and below the other. This means that d_1 is positive and d_2 is negative, and consequently $d_1 = -d_2$. We find, by formula (15), that

$$d_1 = \frac{x + 3y - 3}{\sqrt{10}} \qquad \text{and} \qquad d_2 = \frac{x - 2y + 1}{-\sqrt{5}}$$

We now have

$$\frac{x + 3y - 3}{\sqrt{10}} = -\frac{x - 2y + 1}{-\sqrt{5}}$$

$$\sqrt{5}(x + 3y - 3) = \sqrt{10}(x - 2y + 1)$$

$$x + 3y - 3 = \sqrt{2}(x - 2y + 1)$$

$$(1 - \sqrt{2})x + (3 + 2\sqrt{2})y = 3 + \sqrt{2}$$

EXERCISES

Find the directed distance from the line to the point in Exercises 1 through 8.

1. $12x + 5y + 5 = 0$; $(2, -1)$
2. $4x - 3y = 3$; $(2, -5)$
3. $3x - 4y - 16 = 0$; $(-4, 6)$
4. $5x - 12y = 13$; $(-1, -2)$
5. $x - y + 3 = 0$; $(4, 5)$
6. $3x + y = 20$; $(0, -9)$
7. $x + 4 = 0$; $(-1, -5)$
8. $y - 7 = 0$; $(3, -5)$

Find the distance between the two parallel lines in Exercises 9 through 14.

9. $4x - 3y - 9 = 0$
   $4x - 3y - 24 = 0$
10. $12x + 5y = 13$
    $12x + 5y = 104$
11. $15x - 8y - 34 = 0$
    $15x - 8y + 51 = 0$
12. $x + y + 7 = 0$
    $x + y - 11 = 0$
13. $10x - 24y = 117$
    $5x - 12y = -52$
14. $2x - 3y + 6 = 0$
    $4x - 6y + 5 = 0$

15. Find the equation of the bisector of the pair of acute angles formed by the lines $4x + 2y = 9$ and $2x - y = 8$.

16. Find the equation of the bisector of the acute angles and also the bisector of the obtuse angles formed by the lines $x + 2y - 3 = 0$ and $2x + y - 4 = 0$.

17. Find the equation of the bisector of the acute angles and also of the obtuse angles formed by the lines $3x + 4y - 12 = 0$ and $12x - 4y - 20 = 0$.

18. The vertices of a triangle are at $A(2,4)$, $B(-1,2)$, and $C(7,-1)$. Find the length of altitude from the vertex A and the length of side BC. Then compute the area of the triangle.

19. The vertices of a triangle are at $A(-1,1)$, $B(7,-2)$, and $C(5,3)$. Find the length of one side of the triangle and the length of the altitude to that side, and then compute the area of the triangle.

4. FAMILIES OF LINES

We have expressed equations of lines in various forms. Among these are the equations

$$y = mx + b \qquad \text{and} \qquad \frac{x}{a} + \frac{y}{b} = 1$$

Each of these equations has two constants which have geometrical significance. The constants of the first equation are m and b. When definite values are assigned to these letters, a line is completely determined. Other values for these constants, of course, determine other lines. Thus the quantities m and b

are fixed for any particular line but change from line to line. These letters are called **parameters**. In the second equation a and b would be called the parameters.

A linear equation with one parameter represents lines all with a particular property. For example, the equation $y = 3x + b$ represents a line with slope 3 and y-intercept b. We consider b a parameter which may assume any real value. Since the slope is the same for all values of b, the equation represents a set of parallel lines. The totality of lines thus determined is called a **family** of lines. There are, of course, infinitely many lines in the family. In fact, a line of the family passes through each point of the coordinate plane.

Next, let us consider the linear equation

$$2x - 3y = k$$

This equation represents a line of slope $\frac{2}{3}$ for any particular value of the parameter k. Hence the equation, with k allowed to vary, represents a family of parallel lines with slope $\frac{2}{3}$. Figure 8 shows a few lines of the family corresponding to the indicated values of k.

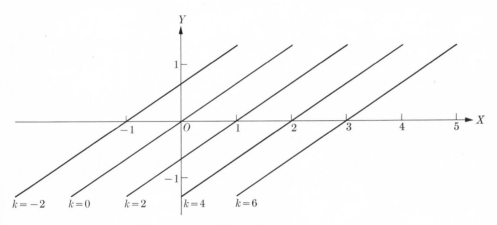

Figure 8

For another illustration of a family of lines, we choose the equation

$$y - 2 = m(x - 4)$$

This is the equation of the family of lines passing through the point $(4,2)$. The family consists of all lines through this point except the vertical line. There is no value of the parameter m which will yield the vertical line. A few lines of the family are drawn in Fig. 9.

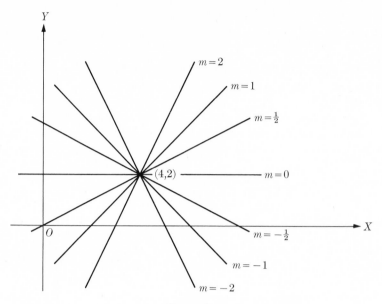

Figure 9

Example 1. Write the equation of the family of lines defined by each of the following conditions:

    a) perpendicular to $3x - 2y = 5$
    b) passing through $(5, -2)$
    c) the product of the intercepts equal to 4

Solution. The following equations are those required:

    a) $2x + 3y = k$
    b) $y + 2 = m(x - 5)$
    c) $\dfrac{x}{a} + \dfrac{y}{4/a} = 1$ or $4x + a^2 y = 4a,\ a \neq 0$

Example 2. Write the equation of the family of lines which are parallel to the line $5x + 12y + 7 = 0$. Find the equations of the members of the family which are 3 units from the point $(2, 1)$.

Solution. Each member of the family $5x + 12y + C = 0$ is parallel to the given line. We seek values of the parameter C which will yield lines 3 units from the point $(2, 1)$, one above the point and the other below the point. Using the formula for the distance from a line to a point, we obtain the equations

$$\frac{5(2) + 12(1) + C}{13} = 3 \quad \text{and} \quad \frac{5(2) + 12(1) + C}{13} = -3$$

The roots are $C = 17$ and $C = -61$. Hence the required equations are

$$5x + 12y + 17 = 0 \quad \text{and} \quad 5x + 12y - 61 = 0$$

The equation of the family of lines passing through the intersection of two given lines can be written readily. To illustrate, we consider the two intersecting lines

$$2x - 3y + 5 = 0 \quad \text{and} \quad 4x + y - 11 = 0$$

From the left members of these equations we form the equation

$$(2x - 3y + 5) + k(4x + y - 11) = 0 \tag{16}$$

where k is a parameter. This equation is of the first degree in x and y for any value of k. Hence it represents a family of lines. Furthermore, each line of the family goes through the intersection of the given lines. We verify this statement by actual substitution. The given lines intersect at $(2,3)$. Then, using these values for x and y, we get

$$(4 - 9 + 5) + k(8 + 3 - 11) = 0$$
$$0 + k(0) = 0$$
$$0 = 0$$

This result demonstrates that Eq. (16) is satisfied by the coordinates $(2,3)$ regardless of the value of k. Hence the equation defines a family of lines passing through the intersection of the given lines.

More generally, let the equations

$$A_1 x + B_1 y + C_1 = 0$$

$$A_2 x + B_2 y + C_2 = 0$$

define two intersecting lines. Then the equation

$$(A_1 x + B_1 y + C_1) + k(A_2 x + B_2 y + C_2) = 0$$

represents a family of lines through the intersection of the given lines. To verify this statement, we first observe that the equation is linear for any value of k. Next we notice that the coordinates of the intersection point reduce each of the parts in parentheses to zero, and hence satisfy the equation for any value of k.

Example 3. Write the equation of the family of lines that pass through the intersection of $2x - y - 1 = 0$ and $3x + 2y - 12 = 0$. Find the member of the family which passes through $(-2, 1)$.

Solution. The equation of the family of lines passing through the intersection of the given lines is

$$(2x - y - 1) + k(3x + 2y - 12) = 0 \tag{17}$$

To find the member of the family passing through $(-2, 1)$, we replace x by -2 and y by 1. This gives

$$(-4 - 1 - 1) + k(-6 + 2 - 12) = 0$$

$$k = -\tfrac{3}{8}$$

Replacing k by $-\tfrac{3}{8}$ in Eq. (17), we have

$$(2x - y - 1) - \tfrac{3}{8}(3x + 2y - 12) = 0$$

or, on simplifying,

$$x - 2y + 4 = 0$$

Example 4. Write the equation of the family of lines through the intersection of $x - 7y + 3 = 0$ and $4x + 2y - 5 = 0$. Find the member of the family which has the slope 3.

Solution. The equation of the family of lines passing through the intersection of the given lines is

$$(x - 7y + 3) + k(4x + 2y - 5) = 0$$

or, collecting terms,

$$(1 + 4k)x + (-7 + 2k)y + 3 - 5k = 0$$

The slope of each member of this family, except for the vertical line, is $-(1 + 4k)/(2k - 7)$. Equating this fraction to the required slope gives

$$-\frac{1 + 4k}{2k - 7} = 3 \quad \text{and} \quad k = 2$$

The member of the system for $k = 2$ is $9x - 3y - 7 = 0$.

EXERCISES

Write the equation of the family of lines possessing the given property in Exercises 1 through 7. In each case assign three values to the parameter and graph the corresponding lines.

1. The lines are parallel to $7x - 4y = 3$.
2. The lines pass through the point $(-3, 4)$.

3. The x intercept is twice the y intercept.

4. The lines are perpendicular to $2x - 4y + 3 = 0$.

5. The y intercept is equal to 4.

6. The sum of the intercepts is equal to 10.

7. The product of the intercepts is 32.

Tell what geometric property is possessed by all lines of each family in Exercises 8 through 16. The letters other than x and y are parameters.

8. $y = mx + 4$ 9. $y = 2x + b$ 10. $9x + 2y = k$

11. $y + 4 = m(x - 3)$ 12. $Ax + 3y = A$ 13. $x - By = 2B$

14. $3x + ay = 3a$ 15. $ax + 2y = 4a$ 16. $ax - ay = 3$

17. In each of Exercises 9, 11, 13, and 15, determine the equation of the line of the family which passes through $(2, -3)$.

18. Write the equation of the family of lines with slope -3, and find the equations of the two lines of the family which are 5 units from the origin.

19. Write the equation of the family of lines parallel to $4x + 3y + 25 = 0$, and find the equations of the two lines which are 6 units from the point $(-2, 3)$.

20. The line $3x - 2y + 1 = 0$ is midway between two parallel lines which are 8 units apart. Find the equations of the two lines.

21. The line $3x - y + 6 = 0$ is midway between two parallel lines which are 4 units apart. Find the equations of the two lines.

In Exercises 22 through 29, find the equation of the line which passes through the intersection of the given pair of lines and satisfies the other given condition.

22. $3x + 2y - 2 = 0$, $x + 5y - 4 = 0$; passes through $(5, 2)$.

23. $5x + 3y + 2 = 0$, $x - y + 2 = 0$; passes through $(-2, -4)$.

24. $4x - 2y + 7 = 0$, $2x + 5y + 4 = 0$; the x intercept is 3.

25. $2x - y - 5 = 0$, $x + y - 4 = 0$; passes through $(0, 0)$.

26. $3x - 4y - 2 = 0$, $3x + 4y + 1 = 0$; the intercepts are equal.

27. $2x - y - 1 = 0$, $3x + 2y - 12 = 0$; the intercepts are equal.

28. $5x + 3y + 2 = 0$, $x - y - 2 = 0$; $m = -3$.

29. $x - 11y = 0$, $3x + y - 5 = 0$; a vertical line.

30. The sides of a triangle are on the lines defined by $2x - 3y + 4 = 0$, $x + y + 3 = 0$, and $5x - 4y - 20 = 0$. Without solving for the vertices, find the equations of the altitudes.

31. The sides of a triangle are on the lines defined by $3x + 5y + 2 = 0$, $x - y - 2 = 0$, and $4x + 2y - 3 = 0$. Without solving for the vertices, find the equations of the altitudes.

5. CIRCLES

We have seen how to write the equation of a line whose position on the coordinate plane is specified. We shall discover that it is equally easy to write the equation of a circle if the location of its center and the radius are known. First, however, we give an explicit definition of a circle.

DEFINITION 1   *A **circle** is the set of all points on a plane that are equidistant from a fixed point on the plane. The fixed point is called the **center**, and the distance from the center to any point of the circle is called the **radius**.*

Let the center of a circle be at the fixed point $C(h,k)$ and let the radius be equal to r. Then if $P(x,y)$ is any point of the circle, the distance from C to P is equal to r (Fig. 10). This condition requires that

$$\sqrt{(x-h)^2 + (y-k)^2} = r$$

and, by squaring,

$$(x-h)^2 + (y-k)^2 = r^2 \tag{18}$$

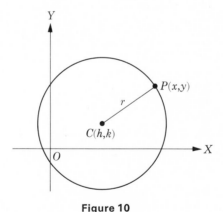

Figure 10

This formula exhibits the coordinates of the center and the length of the radius and, consequently, is sometimes called the **center–radius form** of the equation of a circle.

Conversely, the graph of an equation of the form (18) is a circle with center at (h,k) and radius equal to r. This fact is evident since the equation is satisfied by, and only by, points whose distance from (h,k) is r. Hence it is an easy task to write the equation of a circle whose center and radius are known, or to draw the circle whose equation is expressed in the form (18).

If the center of a circle is at the origin ($h = 0$, $k = 0$) and the radius is r, its equation is

$$x^2 + y^2 = r^2 \qquad (19)$$

Example 1. If the center of a circle is at $(3, -2)$ and the radius is 4, the equation of the circle is

$$(x - 3)^2 + (y + 2)^2 = 16$$

Equation (18) can be presented in another form by squaring the binomials and collecting terms. Thus

$$x^2 - 2hx + h^2 + y^2 - 2ky + k^2 = r^2$$

$$x^2 + y^2 - 2hx - 2ky + h^2 + k^2 - r^2 = 0$$

The last equation is of the form

$$x^2 + y^2 + Dx + Ey + F = 0 \qquad (20)$$

This is called the **general form** of the equation of a circle.

Conversely, an equation of the form (20) can be reduced to the form (18) by the simple expedient of completing the squares in the x terms and the y terms. We illustrate the procedure.

Example 2. Change the equation

$$2x^2 + 2y^2 - 4x + 5y - 6 = 0$$

to the form (18).

Solution. We first divide by 2 to reduce the equation to the general form (20). Thus,

$$x^2 + y^2 - 2x + \tfrac{5}{2}y - 3 = 0$$

Next, leaving spaces for the terms to be added to complete the squares, we have

$$x^2 - 2x \qquad + y^2 + \tfrac{5}{2}y \qquad = 3$$

The numbers 1 and $\frac{25}{16}$ need to be inserted in the empty spaces. Thus, we get

$$(x^2 - 2x + 1) + (y^2 + \tfrac{5}{2}y + \tfrac{25}{16}) = 3 + 1 + \tfrac{25}{16}$$

$$(x - 1)^2 + (y + \tfrac{5}{4})^2 = \tfrac{89}{16}$$

This equation is in the required form, and it reveals that the graph of the given equation is a circle with center at $(1, -\frac{5}{4})$ and radius equal to $\sqrt{89}/4$ (Fig. 11).

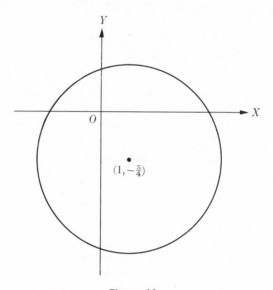

Figure 11

Not all equations of the forms (18), (19), and (20) represent circles. If $r = 0$, Eq. (18) becomes

$$(x - h)^2 + (y - k)^2 = 0$$

This equation is satisfied only by the point (h, k). In this case the graph is sometimes called a **point circle**. The graph of Eq. (20) depends on the constants D, E, and F. By completing the squares, the equation may be presented as

$$\left(x + \frac{D}{2}\right)^2 + \left(y + \frac{E}{2}\right)^2 = \frac{D^2}{4} + \frac{E^2}{4} - F$$

If the right member $D^2/4 + E^2/4 - F$ is positive, the graph is a circle. If the right member is equal to zero, the graph is a point circle. And no real values of x and y satisfy the equation when the right member is negative.

Example 3. What is the graph, if any, of the equation

$$x^2 + y^2 - 4x - 6y + 14 = 0?$$

Solution. Upon completing the squares, we find

$$(x - 2)^2 + (y - 3)^2 = -1$$

Clearly the left member of this equation cannot be negative for any real values of x and y. Hence the equation has no graph.

6. CIRCLES DETERMINED BY GEOMETRIC CONDITIONS

We have seen how to write the equation of a line from certain information which fixes the position of the line in the coordinate plane. We consider now a similar problem concerning the circle. Both the center-radius form and the general form of the equation of a circle will be useful in this connection. There are innumerable geometric conditions which determine a circle. It will be recalled, for example, that a circle can be passed through three points which are not on a straight line. We illustrate this case first.

Example 1. Find the equation of the circle which passes through the points $P(1, -2)$, $Q(5, 4)$, and $R(10, 5)$.

Solution. The equation of the circle can be expressed in the form

$$x^2 + y^2 + Dx + Ey + F = 0$$

Our problem is to find values for D, E, and F so that the equation is satisfied by the coordinates of each of the given points. Hence we substitute for x and y the coordinates of these points. This gives the system

$$1 + 4 + D - 2E + F = 0$$
$$25 + 16 + 5D + 4E + F = 0$$
$$100 + 25 + 10D + 5E + F = 0$$

The solution of these equations is $D = -18$, $E = 6$, and $F = 25$. Therefore the required equation is

$$x^2 + y^2 - 18x + 6y + 25 = 0$$

Alternatively, this problem can be solved by applying the fact that the perpendicular bisectors of two chords of a circle intersect at the center. Thus the equations of the perpendicular bisectors of PQ and QR (Fig. 12) are

$$2x + 3y = 9 \quad \text{and} \quad 5x + y = 42$$

The solution of these equations is $x = 9$, $y = -3$. These are the coordinates of the center. The radius is the distance from the center to either of the given points. Hence the resulting equation, in center-radius form, is

$$(x - 9)^2 + (y + 3)^2 = 65$$

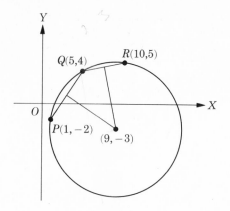

Figure 12

Example 2. A circle is tangent to the line $2x - y + 1 = 0$ at the point $(2,5)$, and the center is on the line $x + y = 9$. Find the equation of the circle.

Solution. The line through $(2,5)$ and perpendicular to the line $2x - y + 1 = 0$ passes through the center of the circle (Fig. 13). The equation of this line is $x + 2y = 12$. Hence the solution of the system

$$x + 2y = 12$$

$$x + y = 9$$

yields the coordinates of the center. Accordingly, the center is at $(6,3)$. The distance from this point to $(2,5)$ is $\sqrt{20}$. The equation of the circle, therefore, is

$$(x - 6)^2 + (y - 3)^2 = 20$$

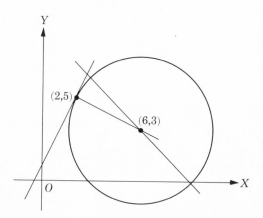

Figure 13

Example 3. A triangle has its sides on the lines $x + 2y - 5 = 0$, $2x - y - 10 = 0$, and $2x + y + 2 = 0$. Find the equation of the circle inscribed in the triangle.

Solution. Referring to Fig. 14, we find that the equations of the bisectors of angles A and B are, respectively, $x = 2$ and $x - 3y = 5$. These bisectors intersect at the point $(2, -1)$. This point is at the distance $\sqrt{5}$ from each side of the triangle. Hence the equation of the circle is

$$(x - 2)^2 + (y + 1)^2 = 5$$

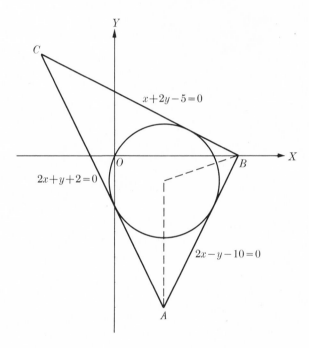

Figure 14

EXERCISES

In Exercises 1 through 16 write the equation of the circle which satisfies the given conditions.

1. Center $(2, -6)$, radius 5

2. Center $(10, 0)$, radius 3

3. Center $(0, 4)$, radius 4

4. Center $(-2, 0)$, radius 7

5. Center $(-12, 5)$, radius 13

6. Center $(4, 3)$, radius 5

7. Center $(\frac{1}{2}, -3)$, radius $\sqrt{11}$

8. Center $(\frac{5}{3}, \frac{1}{3})$, radius $\sqrt{3}$

9. The line segment joining $A(0,0)$ and $B(-8,6)$ is a diameter.

10. The line segment joining $A(5,-1)$ and $B(-7,-5)$ is a diameter.

11. The center is at $(4,2)$, and the circle passes through $(-1,-1)$.

12. The center is at $(-3,1)$, and the circle passes through $(5,-3)$.

13. The circle is tangent to the x axis, and the center is at $(-4,1)$.

14. The circle is tangent to the y axis, and the center is at $(3,5)$.

15. The circle is tangent to the line $3x - 4y = 32$, and the center is at $(0,7)$.

16. The circle is tangent to the line $5x + 12y = 26$, and the center is at the origin.

Reduce each equation 17 through 26 to the center-radius form and construct the circle.

17. $x^2 + y^2 - 6x + 4y - 12 = 0$

18. $x^2 + y^2 - 4x - 12y + 36 = 0$

19. $x^2 + y^2 + 8x + 2y + 1 = 0$

20. $x^2 + y^2 - 10x - 4y - 7 = 0$

21. $x^2 + y^2 - 8x - 6y = 0$

22. $x^2 + y^2 + 10x + 24y = 0$

23. $x^2 + y^2 - 4x + 12y - 8 = 0$

24. $x^2 + y^2 + 3x + 4y = 0$

25. $2x^2 + 2y^2 - 12x + 2y + 1 = 0$

26. $3x^2 + 3y^2 + 6x - 5y = 0$

Determine whether the equations in Exercises 27 through 35 represent a circle, a point, or have no graph.

27. $1 - x^2 - y^2 = 0$

28. $x^2 + y^2 + 1 = 0$

29. $x^2 + y^2 + 2x + 1 = 0$

30. $x^2 + y^2 - 6y = -9$

31. $x^2 + y^2 + x - y = 0$

32. $x^2 + y^2 - 8x + 15 = 0$

33. $x^2 + y^2 + 2x + 10y + 26 = 0$

34. $x^2 + y^2 - 7x - 5y + 40 = 0$

35. $x^2 + y^2 - 3x + 3y + 10 = 0$

Find the equation of the circle described in each of Exercises 36 through 45.

36. The circle is tangent to the line $x + y = 2$ at the point $(4,-2)$, and the center is on the x axis.

37. The circle is tangent to the line $2x - y = 3$ at the point $(2,1)$, and the center is on the y axis.

38. The circle is tangent to the line $3x - 4y = 4$ at the point $(-4,-4)$, and the center is on the line $x + y + 7 = 0$.

39. The circle is tangent to the line $5x - y = 3$ at the point $(2,7)$, and the center is on the line $x + 2y = 19$.

40. The circle is tangent to the line $3x + 4y = 23$ at the point $(5,2)$ and also tangent to the line $4x - 3y + 11 = 0$ at the point $(-2,1)$.

41. The circle is tangent to the line $4x - 3y + 12 = 0$ at the point $(-3,0)$ and also tangent to the line $3x + 4y - 16 = 0$ at the point $(4,1)$.

42. The circle passes through the points $(3,0)$, $(4,2)$, and $(0,1)$.

43. The circle passes through the points $(0,0)$, $(5,0)$, and $(3,3)$.

44. The circle is circumscribed about the triangle whose vertices are $(-3,-1)$, $(4,-2)$, and $(1,2)$.

45. The circle is circumscribed about the triangle whose vertices are $(-2\ 3)$, $(5,2)$, and $(6,-1)$.

46. The sides of a triangle are on the lines $3x - 4y + 8 = 0$, $3x + 4y - 32 = 0$, and $x = 8$. Find the equation of the inscribed circle.

47. The sides of a triangle lie on the lines $3x + y - 5 = 0$, $x - 3y - 1 = 0$, and $x + 3y + 7 = 0$. Find the equation of the circle inscribed in the triangle.

48. The sides of a triangle are on the lines $6x - 7y + 11 = 0$, $2x + 9y + 11 = 0$, and $9x - 2y - 11 = 0$. Find the equation of the circle inscribed in the triangle.

7. FAMILIES OF CIRCLES

In Section 4 we learned how to find the equation of the family of lines passing through the intersection of two lines. The method used there will also serve for finding the equation of families of circles passing through the intersections of two circles. For this purpose, let us consider the equations

$$x^2 + y^2 + D_1x + E_1y + F_1 = 0 \qquad (21)$$

$$x^2 + y^2 + D_2x + E_2y + F_2 = 0 \qquad (22)$$

and, taking k as a parameter, the equation

$$(x^2 + y^2 + D_1x + E_1y + F_1) + k(x^2 + y^2 + D_2x + E_2y + F_2) = 0 \qquad (23)$$

Suppose now that Eqs. (21) and (22) represent circles which intersect in two points. Then if k is a parameter, with k different from -1, Eq. (23) represents a family of circles passing through the intersection points of the given equations. This is true because the coordinates of an intersection point, when substituted for x and y, reduce Eq. (23) to $0 + k0 = 0$. If the given circles are tangent to each other, Eq. (23) represents the family of circles passing through the point of tangency. We shall give examples of circles intersecting at two points and of circles intersecting at only one point.

Example 1. Write the equation of the family of circles all members of which pass through the intersection of the circles C_1 and C_2 represented by the equations

$$C_1: \quad x^2 + y^2 - 6x + 2y + 5 = 0$$
$$C_2: \quad x^2 + y^2 - 12x - 2y + 29 = 0$$

Find the member of the family C_3 which passes through the point $(7,0)$.

Solution. Letting k be a parameter, we express the family of circles by the equation

$$(x^2 + y^2 - 6x + 2y + 5) + k(x^2 + y^2 - 12x - 2y + 29) = 0 \qquad (24)$$

Replacing x by 7 and y by 0 in this equation, we find

$$(49 - 42 + 5) + k(49 - 84 + 29) = 0$$

$$12 + k(-6) = 0$$

$$k = 2$$

For this value of k, Eq. (24) reduces to

$$3x^2 + 3y^2 - 30x - 2y + 63 = 0$$

or, in center-radius form,

$$C_3: \quad (x - 5)^2 + (y - \tfrac{1}{3})^2 = \tfrac{37}{9}$$

Hence the required member of the family of circles has its center at $(5, \tfrac{1}{3})$ and radius equal to $\sqrt{37}/3$, which is approximately equal to 2.03. This circle and the two given circles are constructed in Fig. 15.

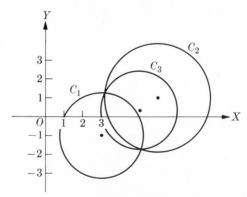

Figure 15

Example 2. Graph the circles C_1 and C_2 whose equations are

$$C_1: \quad x^2 + y^2 - 12x - 9y + 50 = 0$$

$$C_2: \quad x^2 + y^2 - 25 = 0$$

Also graph the member C_3 of the family of circles

$$(x^2 + y^2 - 12x - 9y + 50) + k(x^2 + y^2 - 25) = 0$$

for which $k = 1$.

Solution. Replacing k by 1, we obtain, in center-radius form, the equation

$$C_3: \quad (x - 3)^2 + (y - \tfrac{9}{4})^2 = \tfrac{25}{16}$$

The center of the circle is at $(3, \tfrac{9}{4})$ and the radius is $\tfrac{5}{4}$. As shown in Fig. 16, the given circles intersect in only one point and the third circle passes through the point of tangency.

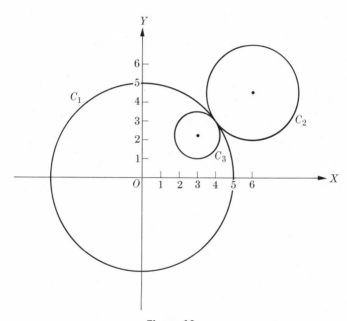

Figure 16

Equation (23), as we saw, represents a circle if k is assigned any real number except $k = -1$. If, however, we let $k = -1$, Eq. (23) reduces to the linear equation

$$(D_1 - D_2)x + (E_1 - E_2)y + F_1 - F_2 = 0$$

The graph of this equation is a straight line called the **radical axis** of the two given circles. If the given circles intersect in two points, the radical axis passes through the intersection points; if the given circles are tangent, the radical axis is tangent to the circles at their point of tangency. If the given circles have no common point, the radical axis is between the circles. In each of the three possibilities, the radical axis is perpendicular to the line joining the centers of the given circles. We leave the proof of this statement to the student.

EXERCISES

1. Draw the radical axis and the circles represented by the equations

$$x^2 + y^2 - 4 = 0 \qquad \text{and} \qquad x^2 + y^2 - 14x + 40 = 0$$

2. Draw the radical axis and the circles represented by the equations

$$x^2 + y^2 - 2x + 4y - 4 = 0 \qquad \text{and} \qquad x^2 + y^2 - 12x - 8y + 36 = 0$$

3. Prove that the radical axis of two circles is perpendicular to the line segment joining the centers of the circles.

4. Construct the graphs of the equations

$$x^2 + y^2 - 4x = 0 \qquad \text{and} \qquad x^2 + y^2 - 4 = 0$$

Find, in center-radius form, the equation of the member of the family of circles

$$(x^2 + y^2 - 4x) + k(x^2 + y^2 - 4) = 0$$

for which $k = 1$. Draw the graph of the equation on the same coordinate system as the given equations.

5. Write the equation of the family of circles represented by the equations

$$x^2 + y^2 - 4x - 1 = 0 \qquad \text{and} \qquad x^2 + y^2 - 10x - 4y - 21 = 0$$

and find the member of the family which passes through the point $(6, 1)$. Construct the graphs of the three circles.

6. Write the equation of the family of circles passing through the intersection points of

$$x^2 + y^2 - 16x - 10y + 24 = 0 \qquad \text{and} \qquad x^2 + y^2 - 4x + 8y - 6 = 0$$

Find the member of the family which passes through the origin. Draw the three circles.

7. Write the equation of the family of circles passing through the intersection of the circles

$$x^2 + y^2 + 2x - 24 = 0 \qquad \text{and} \qquad x^2 + y^2 - 10x - 9y + 39 = 0$$

Find the member of the family for which $k = 1$. Construct the three circles.

8. TRANSLATION OF AXES

The equation of a circle of radius r has the simple form $x^2 + y^2 = r^2$ if the origin of coordinates is at the center of the circle. If the origin is not at the center, the corresponding equation may be expressed in either of the less simple forms (18) or (20), Section 5. This is an illustration of the fact that the simplicity of the equation of a curve depends on the relative positions of the curve and the axes.

Suppose we have a curve in the coordinate plane and the equation of the curve. Let us consider the problem of writing the equation of the same curve with respect to another pair of axes. The process of changing from one pair of axes to another is called a **transformation of coordinates.** The most general transformation is one in which the new axes are not parallel to the old axes and the origins are different. Just now, however, we shall consider transformations in which the new axes are parallel to the original axes and similarly directed. A transformation of this kind is called a **translation of axes.**

The coordinates of each point of the plane are changed under a translation of axes. To see how the coordinates are changed, note Fig. 17. The new axes

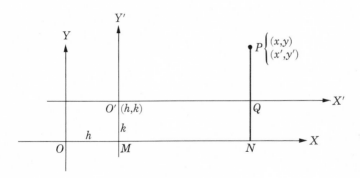

Figure 17

$O'X'$ and $O'Y'$ are parallel, respectively, to the old axes OX and OY. The coordinates of the origin O', referred to the original axes, are denoted by (h, k). Hence the new axes can be obtained by shifting the old axes h units horizontally and k units vertically while keeping their directions unchanged. Let x and y stand for the coordinates of any point P when referred to the old axes, and x' and y' the coordinates of P with respect to the new axes. It is evident from the figure that

$$x = \overrightarrow{ON} = \overrightarrow{OM} + \overrightarrow{O'Q} = h + x'$$

$$y = \overrightarrow{NP} = \overrightarrow{MO'} + \overrightarrow{QP} = k + y'$$

Hence

$$x = x' + h \qquad y = y' + k$$

These formulas give the relations of the old and new coordinates. They hold for all points of the plane, where the new origin O' is any point of the plane. Consequently, the substitutions $x' + h$ for x and $y' + k$ for y in the equation of a curve referred to the original axes yield the equation of the same curve referred to the translated axes.

Example 1. Find the new coordinates of the point $P(4,-2)$ if the origin is moved to $(-2,3)$ by a translation.

Solution. Since we are to find the new coordinates of the given point, we write the translation formulas as $x' = x - h$ and $y' = y - k$. The original coordinates of the given point are $x = -2$, $y = 3$. Making the proper substitutions, we find

$$x' = 4 - (-2) = 6 \qquad y' = -2 - 3 = -5$$

The new coordinates of P are $(6,-5)$. This result can be obtained directly from Fig. 18.

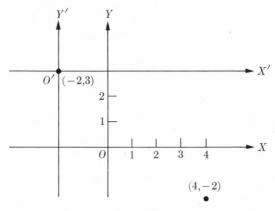

Figure 18

Example 2. Find the new equation of the circle

$$x^2 + y^2 - 6x + 4y - 3 = 0$$

after a translation which moves the origin to the point $(3,-2)$.

Solution. The translation formulas in this case become $x = x' + 3$ and $y = y' - 2$.

These substitutions for x and y in the given equation yield

$$(x' + 3)^2 + (y' - 2)^2 - 6(x' + 3) + 4(y' - 2) - 3 = 0$$

or, by simplification,

$$x'^2 + y'^2 = 16$$

Both sets of axes and the graph are drawn in Fig. 19.

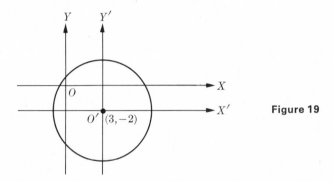

Figure 19

Example 3. Translate the axes so that no first-degree terms will appear in the transformed equation of the circle

$$x^2 + y^2 + 6x - 10y + 12 = 0$$

Solution. We first express the equation in the center-radius form and have

$$(x + 3)^2 + (y - 5)^2 = 22$$

If we choose the new origin at the center of the circle, $(-3,5)$, the translation formulas are $x = x' - 3$ and $y = y' + 5$. These substitutions for x and y give

$$x'^2 + y'^2 = 22$$

as the equation of the circle referred to the translated axes.

EXERCISES

Determine the new coordinates of the points in Exercises 1 through 5 if the axes are translated so that the new origin is at the given point O'. Draw both sets of axes and verify your result from the figure.

1. $(2,3), (-2,3), (-2,-3), (2,-3)$; $O'(1,4)$
2. $(1,7), (2,4), (-3,4), (3,2)$; $O'(3,2)$

3. $(-3, 4), (-1, 5), (-2, -3), (4, 3); O'(1, -3)$

4. $(3, 6), (-3, 2), (4, 7), (-2, -4); O'(-2, -5)$

5. $(3, 6), (-3, 2), (4, 7), (-2, -4); O'(2, -2)$

Find the new equation in Exercises 6 through 13 if the origin is moved to the given point O' by a translation of axes. Draw both sets of axes and the graph.

6. $2x - y - 6 = 0; O'(2, -2)$ 　　　　　 7. $x + 2y - 4 = 0; O'(-2, 3)$

8. $3x - 2y + 5 = 0; O'(-1, 2)$ 　　　　　 9. $x^2 + y^2 - 2x = 0; O'(1, 0)$

10. $x^2 + y^2 + 2x - 4y - 4 = 0; O'(-1, 2)$

11. $x^2 + y^2 - 2x - 6y + 5 = 0; O'(1, 3)$

12. $x^2 + y^2 - 2x + 8y - 3 = 0; O'(1, -4)$

13. $x^2 + y^2 + 16x - 6y + 9 = 0; O'(-8, 3)$

In Exercises 14 through 17 find the point to which the origin must be translated in order that the transformed equation shall have no first-degree term. Give the transformed equation.

14. $x^2 + y^2 - 6x - 4y + 9 = 0$ 　　　　　 15. $x^2 + y^2 + 4x - 2y = 5$

16. $x^2 + y^2 - 10x + 12y = 3$ 　　　　　 17. $4x^2 + y^2 - 16x + 6y = 3$

18. Derive the translation formulas of Section 8 from a figure in which the new origin is in the fourth quadrant.

Chapter 3

Conics

In the preceding chapter, we defined a circle in terms of a set of points. In this chapter, we shall give names to other sets of points, or curves, and derive the corresponding equations. As in the case of a circle, the equations will be of the second degree, or quadratic, in two variables. All the equations, with a few exceptions, define relations some of which are also functions (Chapter 1, Definitions 4 and 5).

The general quadratic equation in x and y may be expressed in the form

$$Ax^2 + Bxy + Cy^2 + Dx + Ey + F = 0 \qquad (1)$$

The graph of a second-degree equation in the coordinates x and y is called a **conic section** or, more simply, a **conic**. This designation comes from the fact the curve can be obtained as the intersection of a right circular cone and a plane.* Conic sections were investigated, particularly by Greek mathematicians, long before analytic methods were introduced. Various properties of conics were discovered, and this phase of geometry received much emphasis. Today the interest in conic sections is enhanced by numerous important theoretical and practical applications.

---

* Let P be a point on a fixed line L. Then the surface composed of all lines through P and forming a constant angle with L is called a **right circular cone**. The line L is the **axis** of the cone, the point P the **vertex**, and each line composing the surface of the cone is called an **element**. The vertex separates the cone into two parts called **nappes**.

Obviously, different kinds of conic sections are possible. A plane not passing the vertex of a cone may cut all the elements of one nappe and make a closed curve (Fig. 1). If the plane is parallel to an element, the intersection extends

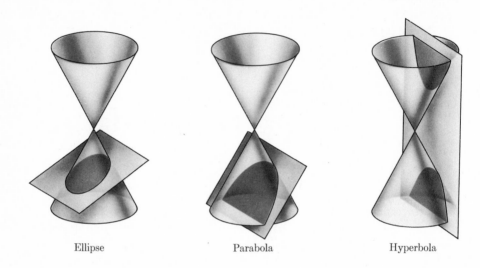

Ellipse   Parabola   Hyperbola

Figure 1

indefinitely far along one nappe but does not cut the other nappe. The plane may cut both nappes and make a section of two parts, each extending indefinitely far along a nappe. In addition to those sections, the plane may pass through the vertex of the cone and determine a point, a line, or two intersecting lines. An intersection of each of these kinds is sometimes called a **degenerate conic.**

1. THE PARABOLA

We now describe and name a curve which, like the circle, is an important concept in mathematics and is used frequently in applied problems.

DEFINITION 1   *A **parabola** is the set of all points in a plane which are equidistant from a fixed point and a fixed line of the plane. The fixed point is called the **focus** and the fixed line the **directrix**.*

In Fig. 2 the point F is the focus and the line D the directrix. The point V, midway between the focus and the directrix, must belong to the parabola (Definition 1). This point is called the **vertex.**

Other points of the parabola can be located in the following way. Draw a line L parallel to the directrix (Fig. 2). With F as a center and radius equal to the distance between the lines D and L, describe arcs cutting L at P and P'. Each of these points, being equidistant from the focus and directrix, is on the parabola. The curve can be sketched by determining a few points in this

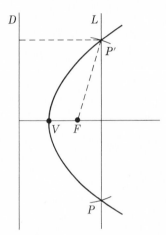

Figure 2

manner. The line VF through the vertex and focus is the perpendicular bisector of PP' and of all other chords similarly drawn. For this reason the line is called the **axis** of the parabola, and the parabola is said to be **symmetric** with respect to its axis.

Although points of a parabola can be located by a direct application of the definition of a parabola, it is easier to obtain them from an equation of the curve. The simplest equation of a parabola can be written if the coordinate axes are placed in a special position relative to the directrix and focus. Let the x axis be on the line through the focus and perpendicular to the directrix and let the origin be at the vertex. Then, choosing $a > 0$, we denote the coordinates of the focus by $F(a,0)$, and the equation of the directrix by $x = -a$ (Fig. 3). Since any point $P(x,y)$ of the parabola is the same distance from the focus and directrix, we have

$$\sqrt{(x - a)^2 + y^2} = x + a$$

Squaring and collecting terms, we obtain the equation

$$y^2 = 4ax$$

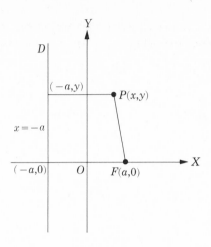

Figure 3

This is the equation of a parabola with the vertex at the origin and focus at $(a,0)$. Since $a > 0$, x may have any positive value or zero, but no negative value. Hence the graph extends indefinitely far into the first and fourth quadrants, and the axis of the parabola is the positive x axis (Fig. 4). It is evident from the equation that the parabola is symmetric with respect to its axis because $y = \pm 2\sqrt{ax}$.

The chord drawn through the focus and perpendicular to the axis of the parabola is given the Latin name **latus rectum**. The length of the latus rectum

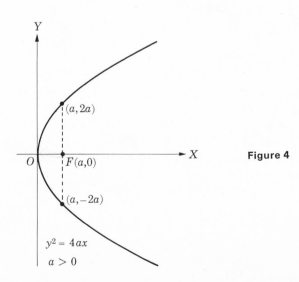

Figure 4

can be determined from the coordinates of its endpoints. By substituting a for x in the equation $y^2 = 4ax$, we find

$$y^2 = 4a^2 \qquad \text{and} \qquad y = \pm 2a$$

Hence the endpoints are $(a, -2a)$ and $(a, 2a)$. This makes the length of the latus rectum equal to $4a$. The vertex and the extremities of the latus rectum are sufficient for drawing a rough sketch of the parabola.

We can, of course, have the focus of a parabola to the left of the origin. For this case, we choose $a < 0$, and let $F(a, 0)$ denote the focus and $x = -a$ the directrix (Fig. 5). Then the positive measurement from a point $P(x, y)$ of the

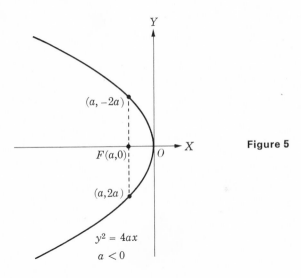

Figure 5

parabola to the directrix is $-a - x$. Hence

$$\sqrt{(x-a)^2 + y^2} = -a - x$$

and this equation, as before, reduces to

$$y^2 = 4ax$$

Since $a < 0$, the variable x can take only negative values and zero, as shown in the figure.

In the preceding discussion, we placed the x axis on the line through the focus and perpendicular to the directrix. By choosing this position for the

y axis, the roles of x and y would be interchanged. Hence the equation of the parabola would then be

$$x^2 = 4ay$$

The graph of this equation when $a > 0$ is in Fig. 6 and when $a < 0$ in Fig. 7.

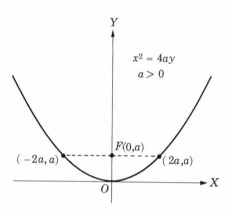

Figure 6

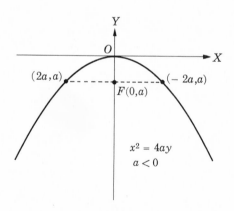

Figure 7

Summarizing, we make the following statements.

THEOREM 1   *The equation of a parabola with vertex at the origin and focus at* $(a, 0)$ *is*

$$\boxed{y^2 = 4ax} \tag{2}$$

The parabola opens to the right if $a > 0$ and opens to the left if $a < 0$.
The equation of a parabola with vertex at the origin and focus at $(0, a)$ is

$$\boxed{x^2 = 4ay} \tag{3}$$

The parabola opens upward if $a > 0$ and opens downward if $a < 0$.

Equations (2) and (3) can be applied to find the equations of parabolas which satisfy specified conditions. We illustrate their use in some examples.

Example 1. Write the equation of the parabola with vertex at the origin and the focus at $(0,4)$.

Solution. Equation (3) applies here. The distance from the vertex to the focus is 4, and hence $a = 4$. Substituting this value for a, we get

$$x^2 = 16y$$

Example 2. A parabola has its vertex at the origin, its axis along the x axis, and passes through the point $(-3, 6)$. Find its equation.

Solution. The equation of the parabola is of the form $y^2 = 4ax$. To determine the value of $4a$, we substitute the coordinates of the given point in this equation. Thus we obtain

$$36 = 4a(-3) \quad \text{and} \quad 4a = -12$$

The required equation is $y^2 = -12x$. The focus is at $(-3, 0)$, and the given point is the upper end of the latus rectum. The graph is constructed in Fig. 8.

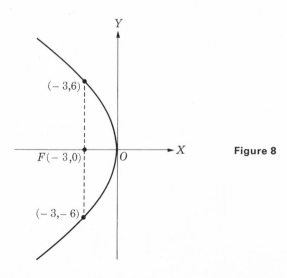

Figure 8

Example 3. The equation of a parabola is $x^2 = -6y$. Find the coordinates of the focus, the equation of the directrix, and the length of the latus rectum.

Solution. The equation is of the form (3), where a is negative. Hence the focus is on the negative y axis and the parabola opens downward. From the equation $4a = -6$, we find $a = -\frac{3}{2}$. Therefore the coordinates of the focus are $(0, -\frac{3}{2})$ and the directrix is $y = \frac{3}{2}$. The length of the latus rectum is equal to the absolute value of $4a$, and in this case is 6. The latus rectum extends 3 units to the left and 3 units to the right of the focus. The graph may be sketched by drawing through the vertex and the ends of the latus rectum. For more accurate graphing a few additional points could be plotted. (See Fig. 9.)

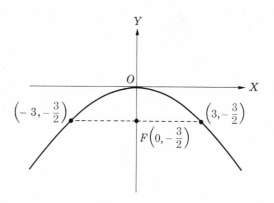

Figure 9

EXERCISES

In Exercises 1 through 9 find the coordinates of the focus, the length of the latus rectum and the coordinates of its endpoints, and the equation of the directrix of each of the given parabolas. Sketch each curve.

1. $y^2 = 8x$ 2. $x^2 = 8y$ 3. $y^2 = -12x$

4. $x^2 = -12y$ 5. $x^2 + 10y = 0$ 6. $5x + 4y^2 = 0$

7. $2y^2 = 9x$ 8. $2y^2 = -5x$ 9. $x^2 - 9y = 0$

Write the equation of the parabola with vertex at the origin which satisfies the given conditions in each of Exercises 10 through 19.

10. Focus at $(2, 0)$ 11. Focus at $(0, 2)$

12. Focus at $(-2, 0)$ 13. Focus at $(0, -2)$

14. Directrix is $x - 4 = 0$ 15. Directrix is $y + 4 = 0$

16. The length of the latus rectum is 6 and the parabola opens to the right.

17. The focus is on the x axis and the parabola passes through the point $(4,3)$.

18. The length of the latus rectum is 10 and the parabola opens upward.

19. The parabola opens to the left and passes through the point $(-2,1)$.

20. On the same set of axes sketch the parabolas $y^2 = 4ax$ for $a = \frac{1}{4}, \frac{1}{2}, 2, 4$, and note the effect that changing the value of a has on the shape of the parabola.

21. A cable suspended from supports which are 200 ft apart has a sag of 50 ft. If the cable hangs in the form of a parabola, find its equation, taking the origin at the lowest point.

22. Find the width of the cable of Exercise 21 at a height 25 ft above the lowest point.

23. Find the equation of the set of all points in the coordinate plane which are equidistant from the point $(0,a)$ and the line $y = -a$, and thus derive Eq. (3) in Chapter 3, Section 1.

2. PARABOLA WITH VERTEX AT (h,k)

We now consider a parabola whose axis is parallel to, but not on, a coordinate axis. In Fig. 10, the vertex is at (h, k) and the focus at $(h + a, k)$. We introduce another pair of axes by a translation to the point (h, k). Since the distance from the vertex to the focus is a, we have at once the equation

$$y'^2 = 4ax'$$

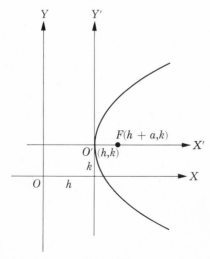

Figure 10

To write the equation of the parabola with respect to the original axes, we apply the translation formulas of Section 8, Chapter 2 and thus obtain

$$(y - k)^2 = 4a(x - h)$$

We observe from this equation, and also from the figure, that when $a > 0$, the factor $x - h$ of the right member must be greater than or equal to zero. Hence the parabola opens to the right. For $a < 0$, the factor $x - h$ must be less than or equal to zero, and therefore the parabola would open to the left. The axis of the parabola is on the line $y - k = 0$. The length of the latus rectum is equal to the absolute value of $4a$, and hence the endpoints can be easily located.

A similar discussion can be made if the axis of a parabola is parallel to the y axis. Consequently, we make the following statements.

THEOREM 2   *The equation of a parabola with vertex at (h, k) and focus at $(h + a, k)$ is*

$$(y - k)^2 = 4a(x - h) \tag{4}$$

The parabola opens to the right if $a > 0$ and opens to the left if $a < 0$.
The equation of a parabola with vertex at (h, k) and focus at $(h, k + a)$ is

$$(x - h)^2 = 4a(y - k) \tag{5}$$

The parabola opens upwards if $a > 0$ and opens downwards if $a < 0$.

Each of Eqs. (4) and (5) is said to be in **standard form**. When $h = 0$ amd $k = 0$, they reduce to the simpler equations of the preceding section. If the equation of a parabola is in standard form, its graph can be quickly sketched, The vertex and the ends of the latus rectum are sufficient for a rough sketch. The plotting of a few additional points would, of course, improve the accuracy.

We note that each of Eqs. (4) and (5) is quadratic in one variable and linear in the other variable. This fact can be expressed more vividly if we perform the indicated squares and transpose terms to obtain the general forms

$$x^2 + Dx + Ey + F = 0 \tag{6}$$

$$y^2 + Dx + Ey + F = 0 \tag{7}$$

Conversely, an equation in the form (6) or (7) can be presented in a standard form, provided $E \neq 0$ in (6) and $D \neq 0$ in (7).

Example 1. Draw the graph of the equation

$$y^2 + 8x - 6y + 25 = 0$$

Solution. The equation represents a parabola because y appears quadratically and x linearly. The graph can be more readily drawn if we first reduce the equation to a standard form. Thus

$$y^2 - 6y + 9 = -8x - 25 + 9$$

$$(y - 3)^2 = -8(x + 2)$$

The vertex is at $(-2, 3)$. Since $4a = -8$ and $a = -2$, the focus is two units to the left of the vertex. The length of the latus rectum, equal to the absolute value of $4a$, is 8. Hence the latus rectum extends 4 units above and below the focus. The graph is constructed in Fig. 11.

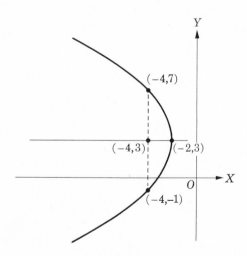

Figure 11

Example 2. A parabola whose axis is parallel to the y axis passes through the points $(1, 1)$, $(2, 2)$, and $(-1, 5)$. Find its equation.

Solution. Since the axis of the parabola is parallel to the y axis, the equation must be quadratic in x and linear in y. Hence we start with the general form

$$x^2 + Dx + Ey + F = 0$$

This equation is to be satisfied by the coordinates of each of the given points. Substituting the coordinates of each point, in turn, we have the system of equations:

$$1 + D + E + F = 0$$

$$4 + 2D + 2E + F = 0$$

$$1 - D + 5E + F = 0$$

The simultaneous solution of these equations is $D = -2$, $E = -1$, and $F = 2$. Hence the equation of the parabola is $x^2 - 2x - y + 2 = 0$.

3. SYMMETRY

We have observed that the axis of a parabola bisects all chords of the parabola which are perpendicular to the axis. For this reason a parabola is said to be **symmetric** with respect to its axis. Many other curves possess the property of symmetry, which leads us to the following discussion.

DEFINITION 2   *Two points A and B are said to be symmetric with respect to a line if the line is the perpendicular bisector of the line segment AB. A curve is symmetric with respect to a line if each of its points is one of a pair of points symmetric with respect to the line.*

Two points A and B are symmetric with respect to a point O if O is the midpoint of the line segment AB. A curve is symmetric with respect to a point O if each of its points is one of a pair of points symmetric with respect to O.

Symmetry of a curve with respect to a coordinate axis or the origin is of special interest. Hence we make the following observations. The points (x, y) and $(x, -y)$ are symmetric with respect to the x axis. Accordingly, a curve is symmetric with respect to the x axis if for each point (x, y) of the curve the point $(x, -y)$ also belongs to the curve. Similarly, a curve is symmetric with respect to the y axis if for each point (x, y) of the curve the point $(-x, y)$ also belongs to the curve. The points (x, y) and $(-x, -y)$ are symmetric with respect to the origin. Hence a curve is symmetric with respect to the origin if for each point (x, y) of the curve the point $(-x, -y)$ also belongs to the curve. (See Fig. 12.)

An equation can be easily tested to determine whether its graph is symmetric with respect to either coordinate axis or the origin. Consider, for example, the equation $x^2 = 4y + 6$. If x is replaced by $-x$, the equation is not altered. This means that if x is given a value and then the negative of that value, the corresponding values of y are the same. Hence for each point (x, y) of the graph there is also the point $(-x, y)$ on the graph. Therefore the graph is symmetric with respect to the y axis. On the other hand, the assigning of equal absolute values to y, one positive and the other negative, leads to different corresponding values of x. Hence the graph is not symmetric with respect to the x axis. Similarly, the graph is not symmetric with respect to the origin.

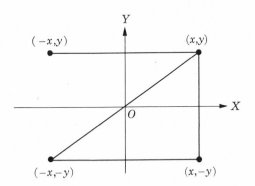

Figure 12

From the definition of symmetry, we formulate the following tests:

1. *If an equation is unchanged when y is replaced by $-y$, then the graph of the equation is symmetric with respect to the x axis.*
2. *If an equation is unchanged when x is replaced by $-x$, then the graph of the equation is symmetric with respect to the y axis.*
3. *If an equation is unchanged when x is replaced by $-x$ and y by $-y$, then the graph of the equation is symmetric with respect to the origin.*

These types of symmetry are illustrated by the equations

$$y^4 - 2y^2 - x = 0 \qquad x^2 - 4y + 3 = 0 \qquad y = x^3$$

The graphs of these equations are symmetric, respectively, with respect to the x axis, the y axis, and the origin. Replacing x by $-x$ and y by $-y$ in the third equation gives $-y = -x^3$, which may be reduced to $y = x^3$.

EXERCISES

Write the equation of the parabola, in standard form, which satisfies the given conditions in Exercises 1 through 10.

1. Vertex $(0, 3)$, focus $(4, 3)$
2. Vertex $(2, 0)$, focus $(2, 2)$
3. Vertex $(2, 3)$, focus $(6, 3)$
4. Vertex $(-3, 1)$, focus $(1, 1)$

5. Vertex $(3, 3)$, focus $(-3, 3)$ 6. Vertex $(5, -2)$, focus $(-2, -2)$

7. Vertex $(-1, -2)$, latus rectum 12; opens downward

8. Vertex $(4, -1)$, latus rectum 8; opens to the right

9. Vertex $(2, 1)$, ends of latus rectum $(-1, -5)$ and $(-1, 7)$

10. Vertex $(3, -2)$, ends of latus rectum $(-2, \frac{1}{2})$ and $(8, \frac{1}{2})$

Express the equations in Exercises 11 through 24 in standard forms. In each case give the coordinates of the vertex, the focus, and the ends of the latus rectum. Sketch the graph.

11. $y^2 - 8x + 8 = 0$ 12. $x^2 - 4y + 8 = 0$

13. $y^2 + 12x - 48 = 0$ 14. $x^2 - 16y - 32 = 0$

15. $x^2 + 4x - 16y + 4 = 0$ 16. $y^2 + 6y - 4x + 9 = 0$

17. $y^2 - 8y + 6x + 16 = 0$ 18. $x^2 + 10x + 20y + 25 = 0$

19. $y^2 + 4y + 8x - 28 = 0$ 20. $x^2 + 2x - 12y + 37 = 0$

21. $x^2 - 8x + 6y - 8 = 0$ 22. $y^2 - 6y + 10x - 1 = 0$

23. $y^2 + 14y - 24x - 119 = 0$ 24. $x^2 - 12x + 16y - 60 = 0$

Find the equation of the parabola in Exercises 25 through 32.

25. Vertex $(-1, -2)$, axis vertical; passes through $(3, 6)$

26. Vertex $(3, -4)$, axis horizontal; passes through $(2, -5)$

27. Axis horizontal; passes through $(1, 1)$, $(1, -3)$, and $(-2, 0)$

28. Axis vertical; passes through $(0, 0)$, $(3, 0)$, and $(-1, 4)$

29. Axis horizontal; passes through $(0, 4)$, $(0, -1)$, and $(6, 1)$

30. Axis vertical; passes through $(-1, 0)$, $(5, 0)$, and $(1, 8)$

31. Axis vertical; passes through $(-1, -3)$, $(1, -2)$, and $(2, 1)$

32. Axis horizontal; passes through $(-1, 1)$, $(3, 4)$, and $(2, -2)$

33. Derive Eq. (4), Section 2, by finding the equation of the set of all points equally distant from the focus $(h + a, k)$ and the directrix $x = h - a$.

34. Derive Eq. (5), Section 2, by finding the equation of the graph of a point equally distant from the focus $(h, k + a)$ and the directrix $y = k - a$.

4. THE ELLIPSE

We obtained second-degree equations for the circle and parabola; hence these curves are conics. We come now to another kind of curve, which, like the circle but unlike the parabola, is a closed curve. The new curve is a conic because its equation, as we shall demonstrate, is of the second degree in x and y.

DEFINITION 3   *An **ellipse** is the set of all points P in a plane such that the sum of the distances of P from two fixed points F' and F of the plane is constant.*

Each of the fixed points is called a **focus** (plural **foci**). Figure 13 shows how the foci can be used in drawing an ellipse. The ends of a string are fastened at the foci F' and F. As the pencil at P moves, with the string taut, the curve traced is an ellipse.

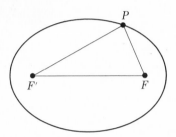

Figure 13

To find the equation of an ellipse, we take the origin of coordinates midway between the foci and one of the coordinate axes on the line through the foci (Fig. 14). We denote the distance between the foci by $2c$, and, accordingly, label the foci as $F'(-c,0)$ and $F(c,0)$. Now if we let the sum of the distances from a point $P(x,y)$ of the ellipse to the foci be $2a$, we obtain

$$PF' + PF = 2a$$

$$\sqrt{(x+c)^2 + y^2} + \sqrt{(x-c)^2 + y^2} = 2a$$

By transposing the second radical, squaring, and simplifying, we get

$$cx - a^2 = -a\sqrt{(x-c)^2 + y^2}$$

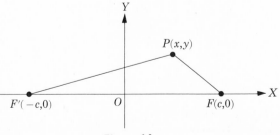

Figure 14

By squaring again and simplifying, we find

$$(a^2 - c^2)x^2 + a^2y^2 = a^2(a^2 - c^2)$$

We observe from the figure that the length of one side of triangle $F'PF$ is $2c$, and the sum of the lengths of the other sides is $2a$. Hence $2a > 2c$, and, consequently, $a^2 - c^2 > 0$. Letting $b^2 = a^2 - c^2$ and dividing by the nonzero quantity a^2b^2, we obtain the final form

$$\boxed{\frac{x^2}{a^2} + \frac{y^2}{b^2} = 1} \qquad (8)$$

We first observe that the graph of Eq. (8) is symmetric with respect to both coordinate axes. When $y = 0$, then $x = \pm a$, and when $x = 0$, $y = \pm b$. Hence the ellipse cuts the x axis at $V'(-a,0)$ and $V(a,0)$ and cuts the y axis at $B'(0,-b)$ and $B(0,b)$ as in Fig. 15. The segment $V'V$ ($=2a$) is called the

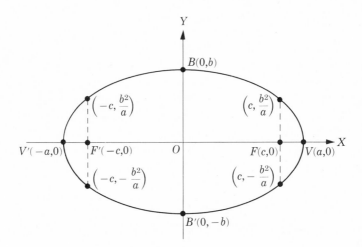

Figure 15

major axis of the ellipse and the segment $B'B(=2b)$ the **minor axis**. The ends of the major axis are called **vertices**. The intersection of the axes of the ellipse is the **center**. The chord through a focus and perpendicular to the major axis is called a **latus rectum**. Substituting $x = c$ in Eq. (8) and using the relation $c^2 = a^2 - b^2$, we find the points $(c, -b^2/a)$ and $(c, b^2/a)$ to be the ends of one latus rectum. The ends of the other latus rectum are at $(-c, -b^2/a)$ and $(-c, b^2/a)$. These results show that the length of each latus rectum is $2b^2/a$. The major axis is longer than the minor axis. This is true because $b^2 = a^2 - c^2 < a^2$, and therefore $b < a$. We note also that the foci are on the major axis. The ellipse and the important points are shown in Fig. 15.

If we take the foci of an ellipse on the y axis at $(0,-c)$ and $(0,c)$, we can obtain, by steps similar to those above, the equation

$$\frac{y^2}{a^2} + \frac{x^2}{b^2} = 1 \tag{9}$$

In this case the major axis is on the y axis and the minor axis on the x axis (Fig. 16).

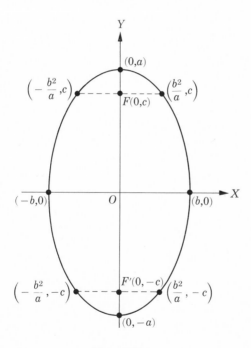

Figure 16

5. EXTENT OF ELLIPSE

From the definition of an ellipse, we know that its points do not extend indefinitely far from the foci. The fact that an ellipse is limited in extent can be deduced from its equation. Thus solving Eq. (8) for x and y in turn, we get

$$x = \pm \frac{a}{b}\sqrt{b^2 - y^2} \quad \text{and} \quad y = \pm \frac{b}{a}\sqrt{a^2 - x^2}$$

These equations reveal that y^2 must not exceed b^2, and x^2 must not exceed a^2. In other words, the permissible values are

$$-b \leqq y \leqq b \quad \text{and} \quad a \leqq x \leqq a.$$

Hence no point of the ellipse is outside the rectangle formed by the horizontal lines $y = -b$, $y = b$ and the vertical lines $x = -a$, $x = a$.

In many equations the extent of the graph can be readily determined by solving for each variable in terms of the other. The concept of extent, as well as that of symmetry, is often an aid in drawing a graph.

Example 1. Find the equation of an ellipse with foci at $(0, \pm 4)$ and a vertex at $(0, 6)$.

Solution. The location of the foci shows that the center of the ellipse is at the origin, that $c = 4$, and that the desired equation may be expressed in the form (9). The given vertex, 6 units from the origin, makes $a = 6$. Using the relation $b^2 = a^2 - c^2$, we find $b^2 = 20$. Hence we obtain the equation

$$\frac{y^2}{36} + \frac{x^2}{20} = 1$$

Example 2. Sketch the ellipse $9x^2 + 25y^2 = 225$.

Solution. Dividing by 225 yields the form

$$\frac{x^2}{25} + \frac{y^2}{9} = 1$$

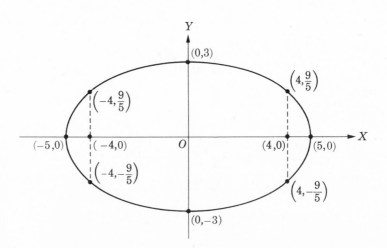

Figure 17

Since the denominator of x^2 is greater than the denominator of y^2, the major axis is along the x axis. We see also that $a^2 = 25$, $b^2 = 9$, and $c = \sqrt{a^2 - b^2} = 4$. Hence the vertices are at $(\pm 5, 0)$, the ends of the minor axis at $(0, \pm 3)$, and the foci at $(\pm 4, 0)$. The length of a latus rectum is $2b^2/a = \frac{18}{5}$. The locations of the ends of the axes and the ends of each latus rectum are sufficient for making a sketch of the ellipse. Figure 17 shows the curve with several important points indicated.

6. FOCUS-DIRECTRIX PROPERTY OF AN ELLIPSE

We defined a parabola in terms of a focus and directrix, but we made no use of a directrix in defining an ellipse. It turns out, however, that an ellipse has a directrix. In deriving the equation of an ellipse (Section 4), we arrived at the equation

$$-a\sqrt{(x-c)^2 + y^2} = cx - a^2$$

This equation implies the existence of a directrix which is evident after a slight modification. Thus, dividing by $-a$ and factoring the right member, we obtain

$$\sqrt{(x-c)^2 + y^2} = \frac{c}{a}\left(\frac{a^2}{c} - x\right)$$

The left member of this equation is the distance from a point (x, y) of the ellipse to the focus $(c, 0)$. The factor $a^2/c - x$ of the right member is the distance from the point of the ellipse to the line $x = a^2/c$ (Fig. 18), and the factor c/a

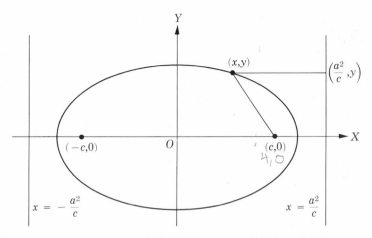

Figure 18

is a constant between zero and unity. Hence we have a proof, based on our definition of an ellipse, of the following theorem.

THEOREM 3   *An ellipse is the set of points in a plane such that the distance of each point of the set from a fixed point of the plane is equal to a constant (between 0 and 1) times its distance from a fixed line of the plane.*

Sometimes an ellipse is defined in terms of a focus and directrix. When so defined, it is possible to prove the statement contained in our definition of an ellipse (Definition 3, Section 4).

The line $x = a^2/c$ is the directrix corresponding to the focus $(c, 0)$. It would be easy to show that the point $(-c, 0)$ and the line $x = -a^2/c$ constitute another focus and directrix. This fact, however, is geometrically evident from considerations of symmetry.

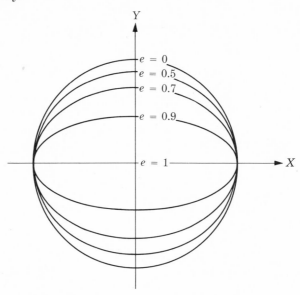

Figure 19

We wish to discuss further the quantity c/a. This ratio is called the **eccentricity** e of the ellipse. The shape of an ellipse depends on the value of its eccentricity. Suppose, for example, we visualize an ellipse in which the major axis remains constant while e starts at zero and approaches unity. If $e = 0$, the equations $e = c/a$ and $b^2 = a^2 - c^2$ show that $c = 0$ and $a = b$. The two foci are then coincident at the center, and the ellipse is a circle. As e increases, the foci separate, each receding from the center, and b decreases. As e approaches 1, c approaches a, and b approaches 0. Hence the ellipse, starting as a circle, becomes narrow and narrower. If $e = 1$, or $c = a$, then $b = 0$. Equation (8),

Section 4, would not then apply because b^2 is a denominator. But, if $e = 1$, the definition of an ellipse requires the graph to be the line segment connecting the foci.

Summarizing, we have an actual ellipse if e is between 0 and 1. When e is slightly more than 0, the ellipse is somewhat like a circle; when e is slightly less than 1, the ellipse is relatively long and narrow.

Ellipses with the same major axis and different eccentricities are constructed in Fig. 19.

7. ELLIPSE WITH CENTER AT (h,k)

If the axes of an ellipse are parallel to the coordinate axes and the center is at (h, k), we can obtain its equation by applying the translation formulas of Section 8, Chapter 2. We draw a new pair of coordinate axes along the axes of the ellipse (Fig. 20). The equation of the ellipse referred to the new axes is

$$\frac{x'^2}{a^2} + \frac{y'^2}{b^2} = 1$$

The substitutions $x' = x - h$ and $y' = y - k$ yield

$$\frac{(x - h)^2}{a^2} + \frac{(y - k)^2}{b^2} = 1 \tag{10}$$

Similarly, when the major axis is parallel to the y-axis, we have

$$\frac{(y - k)^2}{a^2} + \frac{(x - h)^2}{b^2} = 1 \tag{11}$$

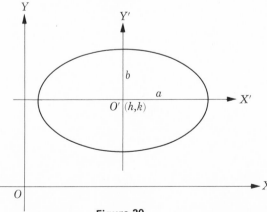

Figure 20

These are the **standard forms** of equations of ellipses. They reduce to Eqs. (8) and (9), Section 4, when $h = k = 0$. The quantities a, b, and c have the same meaning whether or not the center of the ellipse is at the origin. Therefore constructing the graph of an equation in either of the forms of Eqs. (10) or (11) presents no greater difficulty than drawing the graph of one of the simpler Eqs. (8) or (9).

Example 1. Find the equation of the ellipse with foci at $(4,-2)$ and $(10,-2)$, and a vertex at $(12,-2)$.

Solution. The center, midway between the foci, is at $(7,-2)$. The distance between the foci is 6 and the given vertex is 5 units from the center; hence $c = 3$ and $a = 5$. Then $b^2 = a^2 - c^2 = 16$. Since the major axis is parallel to the x axis, we substitute in Eq. (10) and get

$$\frac{(x-7)^2}{25} + \frac{(y+2)^2}{16} = 1$$

The graph of this ellipse is constructed in Fig. 21.

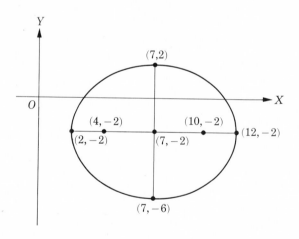

Figure 21

Example 2. Reduce, to standard form, the equation

$$4y^2 + 9x^2 - 24y - 72x + 144 = 0$$

Solution. The essential steps in the process follow.

$$4(y^2 - 6y) + 9(x^2 - 8x) = -144$$

$$4(y^2 - 6y + 9) + 9(x^2 - 8x + 16) = -144 + 36 + 144$$

$$4(y - 3)^2 + 9(x - 4)^2 = 36$$

$$\frac{(y - 3)^2}{9} + \frac{(x - 4)^2}{4} = 1$$

The coordinates of the center are $(4, 3)$; $a = 3$, $b = 2$, and

$$c = \sqrt{a^2 - b^2} = \sqrt{9 - 4} = \sqrt{5}.$$

The vertices are at $(4, 0)$ and $(4, 6)$, and the ends of the minor axis are at $(2, 3)$ and $(6, 3)$. The coordinates of the foci are $(4, 3 - \sqrt{5})$ and $(4, 3 + \sqrt{5})$. The graph is constructed in Fig. 22.

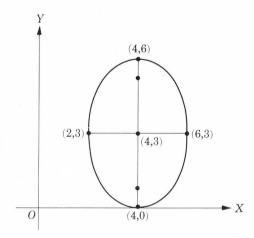

Figure 22

EXERCISES

In Exercises 1 through 12 find the coordinates of the foci, the ends of the major and minor axes, and the ends of each latus rectum. Sketch each curve.

1. $\dfrac{x^2}{25} + \dfrac{y^2}{9} = 1$　　　　　　　　　　2. $\dfrac{x^2}{169} + \dfrac{y^2}{25} = 1$

3. $\dfrac{y^2}{25} + \dfrac{x^2}{16} = 1$　　　　　　　　　　4. $\dfrac{y^2}{169} + \dfrac{x^2}{144} = 1$

$b^2 = a^2 - c^2$

5. $25x^2 + 49y^2 = 1225$ 6. $16x^2 + y^2 = 16$

7. $x^2 + 16y^2 = 16$ 8. $3x^2 + 2y^2 = 18$

9. $\dfrac{(x-2)^2}{16} + \dfrac{(y-3)^2}{9} = 1$ 10. $\dfrac{(x-2)^2}{36} + \dfrac{(y-3)^2}{9} = 1$

11. $\dfrac{(y+5)^2}{169} + \dfrac{(x-5)^2}{49} = 1$ 12. $\dfrac{(y+4)^2}{36} + \dfrac{(x-7)^2}{16} = 1$

Reduce the equations in Exercises 13 through 19 to standard forms. In each, find the coordinates of the center, the foci, and the ends of the major and minor axes. Sketch each ellipse.

13. $16x^2 + 25y^2 - 160x - 200y + 400 = 0$

14. $x^2 + 4y^2 - 6x - 16y + 21 = 0$

15. $16x^2 + 4y^2 - 32x + 16y - 32 = 0$

16. $3x^2 + 2y^2 - 24x + 12y + 60 = 0$

17. $4x^2 + 8y^2 + 4x + 24y - 13 = 0$

18. $25(x+1)^2 + 169(y-2)^2 = 4{,}225$

19. $225(x+2)^2 + 289(y+3)^2 = 65{,}025$

Write the equation of the ellipse which satisfies the given conditions in each of Exercises 20 through 29.

20. Center $(0,0)$, one vertex $(0,-7)$, one end of minor axis $(5,0)$.

21. Center $(0,0)$, one vertex $(3,0)$, one end of minor axis $(0,2)$.

22. Foci $(-3,0)$, and $(3,0)$, vertices $(-5,0)$ and $(5,0)$.

23. Foci $(-5,0)$ and $(5,0)$, length of minor axis 8.

24. Foci $(0,-8)$ and $(0,8)$, length of major axis 34.

25. Vertices $(-5,0)$ and $(5,0)$, length of latus rectum 8/5.

26. Center $(-2,2)$, vertex $(-2,6)$, one end of minor axis $(0,2)$.

27. Foci $(-4,2)$ and $(4,2)$, major axis 10.

28. Focus $(2,3)$, vertex $(2,6)$, minor axis 6.

29. Center $(5,4)$, major axis 16, minor axis 10.

30. The perimeter of a triangle is 14, and the points $(-3,0)$ and $(3,0)$ are two of the vertices. Find the equation of the graph of the third vertex.

31. Find the equation of the graph of the midpoints of the ordinates of the circle $x^2 + y^2 = 49$.

32. The ordinates of a curve are k times the ordinates of the circle $x^2 + y^2 = a^2$. Show that the curve is an ellipse if k is a positive number different from 1.

33. The earth's orbit is an ellipse with the sun at one focus. The length of the major axis is 186,000,000 mi, and the eccentricity is 0.0167. Find the distances from the ends of the major axis to the sun. These are the greatest and least distances from the earth to the sun.

34. The moon's orbit is an ellipse with the earth at one focus. The length of the major axis is 478,000 mi, and the eccentricity is 0.0549. Find the greatest and least distances from the earth to the moon.

35. The arch of an underpass is a semiellipse 60 ft wide and 20 ft high. Find the clearance at the edge of a lane if the edge is 20 ft from the middle.

36. A point moves in the coordinate plane so that the sum of its distances from $(2, 2)$ and $(6, 2)$ is equal to 16. Find the equation of the graph of the point.

37. The perimeter of a triangle is 20, and the points $(2, -3)$ and $(2, 3)$ are two of the vertices. Find the equation of the graph of the third vertex.

38. Find the equation of the graph of a point $P(x, y)$ which moves so that its distance from $(-5, 0)$ is one-half its distance from the line $x = -20$.

39. Find the equation of the graph of a point which moves so that its distance from $(4, 0)$ is equal to two-thirds its distance from the line $x = 9$.

40. Derive Eq. (9), Chapter 3, Section 4, and Eq. (11), Chapter 3, Section 7.

41. Prove that a curve which is symmetric with respect to both coordinate axes is symmetric with respect to the origin.

42. Prove that a curve which is symmetric with respect to a coordinate axis and the origin is symmetric with respect to the other coordinate axis.

43. If p is a parameter of positive numbers, show that all members of the family of ellipses

$$\frac{x^2}{a^2 + p} + \frac{y^2}{b^2 + p} = 1$$

have the same foci.

44. A family of ellipses with the same foci is represented by the equation

$$\frac{x^2}{15 + p} + \frac{y^2}{p} = 1 \qquad p > 0$$

Find the coordinates of the ends of the axes of the ellipses when $p = 1, 10, 21$, and 34. Sketch the four ellipses.

45. Write an equation of the family of ellipses whose common foci are $(-6, 0)$ and $(6, 0)$. Find the member of the family which passes through $(9, 0)$.

8. THE HYPERBOLA

The third type of conic which we shall consider is the hyperbola. The equations of hyperbolas resemble those of ellipses but the properties of these two kinds of conics differ considerably in some respects.

> DEFINITION 4   *A **hyperbola** is the set of points in a plane such that the difference of the distances of each point of the set from two fixed points (foci) in the plane is constant.*

To derive the equation of a hyperbola, we take the origin midway between the foci and a coordinate axis on the line through the foci (Fig. 23). We denote

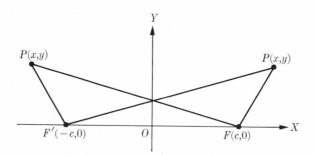

Figure 23

the foci by $F'(-c,0)$ and $F(c,0)$ and the difference of the distances between a point of the hyperbola and the foci by the positive constant $2a$. Then, from the above definition.

$$|F'P| - |FP| = 2a$$

or

$$|F'P| - |FP| = -2a$$

depending on whether the point $P(x,y)$ of the hyperbola is to the right or left of the y-axis. We combine these equations by writing

$$|F'P| - |FP| = \pm 2a$$

or

$$\sqrt{(x+c)^2 + y^2} - \sqrt{(x-c)^2 + y^2} = \pm 2a$$

Now, transposing the second radical, squaring, and simplifying, we obtain

$$cx - a^2 = \pm a\sqrt{(x-c)^2 + y^2}$$

Squaring the members of this equation and simplifying, we get

$$(c^2 - a^2)x^2 - a^2 y^2 = a^2(c^2 - a^2)$$

Then letting $b^2 = c^2 - a^2$ and dividing by $a^2 b^2$, we have

$$\frac{x^2}{a^2} - \frac{y^2}{b^2} = 1 \tag{12}$$

The division is allowable if $c^2 - a^2 \neq 0$. In Fig. 23, the length of one side of triangle $F'PF$ is $2c$ and the difference of the other two sides is $2a$. Hence

$$c > a \qquad \text{and} \qquad c^2 - a^2 > 0$$

    The graph of Eq. (12) is symmetric with respect to the coordinate axes. The permissible values of x and y become evident where each is expressed in terms of the other. Thus we get

$$x = \pm \frac{a}{b} \sqrt{b^2 + y^2} \qquad \text{and} \qquad y = \pm \frac{b}{a} \sqrt{x^2 - a^2}$$

We see from the first of these equations that y may have any real value and from the second that x may have any real value except those for which $x^2 < a^2$. Hence the hyperbola extends indefinitely far from the axes in each quadrant. But there is no part of the graph between the line $x = -a$ and the line $x = a$. Accordingly, the hyperbola consists of two separate parts, or **branches** (Fig. 24). The points $V'(-a,0)$ and $V(a,0)$ are called **vertices**, and the segment

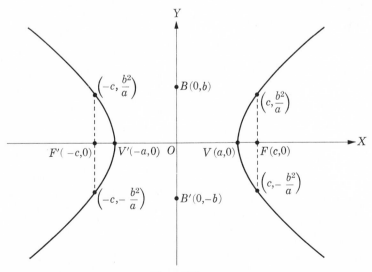

Figure 24

$V'V$ is called the **transverse axis**. The segment from $B'(0,-b)$ to $B(0,b)$ is called the **conjugate axis**. Although the conjugate axis has no point in common with the hyperbola, it has an important relation to the curve, as we shall discover. The intersection of the axes is the **center**. The chord through a focus and perpendicular to the transverse axis is called a **latus rectum**. By substituting $x = c$ in Eq. (12) and using the relation $c^2 = a^2 + b^2$, we find the extremities of a latus rectum to be $(c,-b^2/a)$ and $(c,b^2/a)$. Hence the length is $2b^2/a$.

We note that the meanings of the three positive quantities a, b, and c appearing here are analogous to their meanings when used with an ellipse. The quantity a is the distance from the center of the hyperbola to a vertex, b is the distance from the center to an end of the conjugate axis, and c is the distance from the center to a focus. But the relations of a, b, and c for an ellipse are not the same as for a hyperbola. For an ellipse, $a > c$, $c^2 = a^2 - b^2$, and $a > b$. For a hyperbola, $c > a$, $c^2 = a^2 + b^2$, and no restriction is placed on the relative values of a and b.

If the foci are on the y axis at $F'(0,-c)$ and $F(0,c)$, the equation of the hyperbola is

$$\frac{y^2}{a^2} - \frac{x^2}{b^2} = 1 \tag{13}$$

The vertices are at $V'(0,-a)$ and $V(0,a)$, and the relations among a, b, and c are unchanged.

The generalized equations of hyperbolas with axes parallel to the coordinate axes and centers at (h,k) are

$$\frac{(x-h)^2}{a^2} - \frac{(y-k)^2}{b^2} = 1 \tag{14}$$

$$\frac{(y-k)^2}{a^2} - \frac{(x-h)^2}{b^2} = 1 \tag{15}$$

Equations (12) through (15) are said to be in **standard forms**. The meanings and relations of a, b, and c are the same in all the equations.

9. THE ASYMPTOTES OF A HYPERBOLA

Unlike the other conics, a hyperbola has associated with it two lines bearing an important relation to the curve. These lines are the extended diagonals of

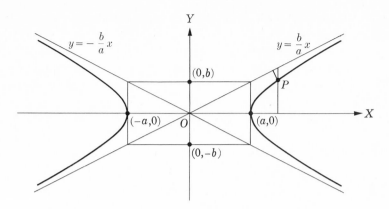

Figure 25

the rectangle in Fig. 25. One pair of sides of the rectangle pass through the vertices and are perpendicular to the transverse axis. The other pair pass through the ends of the conjugate axis. Suppose we consider the extended diagonal and the part of the hyperbola in the first quadrant. The equations of the diagonal and this part of the hyperbola are, respectively,

$$y = \frac{b}{a}\,x \qquad \text{and} \qquad y = \frac{b}{a}\,\sqrt{x^2 - a^2}$$

We see that for any $x > a$ the ordinate of the hyperbola is less than the ordinate of the line. If, however, x is many times as large as a, the corresponding ordinates are almost equal. This may be seen more convincingly by examining the difference of the two ordinates. Thus, by subtracting, and then multiplying the numerator and denominator of the resulting fraction by $x + \sqrt{x^2 - a^2}$, we get

$$\frac{b}{a}\,x - \frac{b}{a}\,\sqrt{x^2 - a^2} = \frac{b(x - \sqrt{x^2 - a^2})}{a}$$

$$= \frac{b(x - \sqrt{x^2 - a^2})(x + \sqrt{x^2 - a^2})}{a(x + \sqrt{x^2 - a^2})}$$

$$= \frac{ab}{x + \sqrt{x^2 - a^2}}$$

The numerator of the last fraction is constant. The denominator, however, increases as x increases. In fact, we can make the denominator as large as we please by taking a sufficiently large value for x. This means that the fraction, which is the difference of the ordinates of the line and the hyperbola, gets closer and closer to zero as x gets larger and larger. The perpendicular distance from a point P of the hyperbola to the line is less than the fraction. Consequently, we can make the perpendicular distance, though never zero, as near zero as we please by taking x sufficiently large. When the perpendicular distance from a line to a curve approaches zero as the curve extends indefinitely far from the origin, the line is said to be an **asymptote** of the curve. From considerations of symmetry, we conclude that each extended diagonal is an asymptote of the curve. Hence the equations of the asymptotes of the hyperbola represented by Eq. (12) are

$$y = \frac{b}{a} x \qquad \text{and} \qquad y = -\frac{b}{a} x$$

Similarly, the equations of the asymptotes associated with Eq. (13) are

$$y = \frac{a}{b} x \qquad \text{and} \qquad y = -\frac{a}{b} x$$

We observe that for each of the hyperbolas, (12) through (15), the equations of the asymptotes may be obtained by factoring the left member and equating each factor to zero.

The asymptotes of a hyperbola are helpful in sketching a hyperbola. A rough drawing can be made from the associated rectangle and its extended diagonals. The accuracy may be improved considerably, however, by plotting the endpoints of each latus rectum.

If $a = b$, the associated rectangle is a square and the asymptotes are perpendicular to each other. For this case the hyperbola is said to be **equilateral** because its axes are equal, or is said to be **rectangular** because its asymptotes intersect at right angles.

The ratio c/a is called the **eccentricity** e of the hyperbola. The angle of intersection of the asymptotes, and therefore the shape of the hyperbola, depends on the value of e. Since $c > a$, the value of e is greater than 1. If c is just slightly greater than a, so that e is near 1, the relation $c^2 = a^2 + b^2$ shows that b is small compared with a. Then the asymptotes make a pair of small angles. The branches of the hyperbola, enclosed by small angles, diverge slowly. If e increases, the branches are enclosed by larger angles. And the angles can be made near $180°$ by taking large values for e.

Having found that the eccentricity of an ellipse is a number between 0 and 1 and that the eccentricity of a hyperbola is greater than 1, we naturally ask if there is a conic whose eccentricity is equal to 1. To have an answer, we recall that any point of a parabola is equally distant from the directrix and the focus (Definition 1). The ratio of these distances is 1, and consequently this value is the eccentricity of a parabola.

Example 1. Sketch the curve $36x^2 - 64y^2 = 2304$.

Solution. We divide by 2304 and reduce the equation to

$$\frac{x^2}{64} - \frac{y^2}{36} = 1$$

The graph is a hyperbola in which $a = 8$, $b = 6$, and $c = \sqrt{a^2 + b^2} = 10$. The vertices therefore are $(\pm 8, 0)$ and the foci $(\pm 10, 0)$. Each latus rectum has a length of $2b^2/a = 9$. The equations of the asymptotes are $3x - 4y = 0$ and $3x + 4y = 0$. From this information the hyperbola can be drawn (Fig. 26).

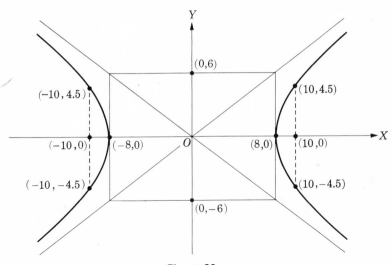

Figure 26

Example 2. Draw the graph of $12y^2 - 4x^2 + 72y + 16x + 44 = 0$.

Solution. We first reduce the equation to a standard form. Thus

$$12(y^2 + 6y + 9) - 4(x^2 - 4x + 4) = -44 + 108 - 16$$

$$12(y + 3)^2 - 4(x - 2)^2 = 48$$

$$\frac{(y + 3)^2}{4} - \frac{(x - 2)^2}{12} = 1 \tag{16}$$

We see now that $a = 2$, $b = 2\sqrt{3}$, $c = \sqrt{4 + 12} = 4$, and that the center of
the hyperbola is at $(2,-3)$. Hence the ends of the transverse axis are at $(2,-5)$
and $(2,-1)$, and the ends of the conjugate axis are at $(2 - 2\sqrt{2},-3)$ and
$(2 + 2\sqrt{2},-3)$. The sides of the associated rectangle pass through these points
(Fig. 27). The extended diagonals of the rectangle are the asymptotes. The

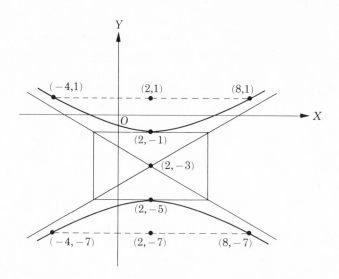

Figure 27

coordinates of the foci, 4 units from the center, are $(2,-7)$ and $(2,1)$. The
length of each latus rectum is $2b^2/a = 12$, and therefore the ends of each latus
rectum are 6 units from a focus. The vertices of the hyperbola, the ends of each
latus rectum, and the asymptotes are sufficient for drawing a reasonably
accurate graph. The equations of the asymptotes, though not needed for their
drawing, are

$$\frac{y + 3}{2} + \frac{x - 2}{2\sqrt{3}} = 0 \quad \text{and} \quad \frac{y + 3}{2} - \frac{x - 2}{2\sqrt{3}} = 0$$

We obtain these equations by factoring the left member of (16) and equating
each factor to zero.

10. APPLICATIONS OF CONICS

Many examples of conics have been discovered in natural phenomena, and
important applications of them abound in engineering and industry.

A projectile, for example, a ball or bullet, travels in a path which is approximately a parabola. The paths of some comets are nearly parabolic. Cables of some suspension bridges hang in the form of a parabola. The surface generated by revolving a parabola about its axis is called a paraboloid of revolution. A reflecting surface in this form has the property that light emanating at the focus is reflected in the direction of the axis. This kind of surface is used in headlights, in some telescopes, and in devices to reflect sound waves. A comparatively recent application of parabolic metal surfaces is found in radar and other microwave equipment. The surfaces reflect radio waves in the same way that light is reflected and are used in directing outgoing beams and also in receiving incoming beams.

The planets have elliptic paths with the sun at a focus. Much use is made of semi-elliptic springs and elliptic-shaped gears. A surface of the form generated by revolving an ellipse about its major axis is so shaped that sound waves emanating at one focus are reflected to arrive at the other focus. This principle is illustrated in whispering galleries and other buildings.

A very interesting and important application of the hyperbola is that of locating the place from which a sound, such as gunfire, emanates. From the difference in the times at which the sound reaches two listening posts, the difference between distances of the posts from the gun can be determined. Then the gun is known to be located on a branch of a hyperbola of which the posts are foci. The position of the gun on this curve can be found by the use of a third listening post. Either of the two posts and the third are foci of a branch of another hyperbola on which the gun is located. Hence the gun is at the intersection of the two branches.

The principle used in finding the location of a gun is also employed by a radar-equipped airplane to determine its location. In this case the plane receives signals from three stations of known locations.

EXERCISES

In Exercises 1 through 10 find the coordinates of the vertices and foci, the length of each latus rectum, and the equations of the asymptotes. Draw the asymptotes and sketch each hyperbola.

1. $\dfrac{x^2}{25} - \dfrac{y^2}{9} = 1$ 2. $\dfrac{y^2}{36} - \dfrac{x^2}{64} = 1$

3. $\dfrac{x^2}{16} - \dfrac{y^2}{4} = 1$ 4. $\dfrac{y^2}{16} - \dfrac{x^2}{4} = 1$

5. $y^2 - x^2 = 49$ 6. $x^2 - y^2 = 36$

7. $\dfrac{y^2}{36} - \dfrac{(x-2)^2}{64} = 1$

8. $\dfrac{x^2}{49} - \dfrac{(y-6)^2}{25} = 1$

9. $\dfrac{(x+3)^2}{16} - \dfrac{(y-2)^2}{9} = 1$

10. $\dfrac{(x-1)^2}{25} - \dfrac{(y+2)^2}{25} = 1$

Reduce each equation, Exercises 11 through 15, to standard form. In each find the coordinates of the center, the vertices, and the foci. Draw the asymptotes and the graph of each equation.

11. $9x^2 - 4y^2 - 36x + 16y - 16 = 0$

12. $3x^2 - 2y^2 - 4y - 26 = 0$

13. $9x^2 - 4y^2 - 90x + 189 = 0$

14. $x^2 - 2y^2 - 6x - 4y + 5 = 0$

15. $49y^2 - 4x^2 - 98y + 48x - 291 = 0$

In each of Exercises 16 through 23 find the equation of the hyperbola which satisfies the given conditions.

16. Center $(0,0)$, transverse axis along the x axis, a focus at $(8,0)$, a vertex at $(4,0)$.

17. Center $(0,0)$, transverse axis along the x axis, a focus at $(5,0)$, transverse axis $= 6$.

18. Center $(0,0)$, transverse axis along the x axis, transverse axis 10, latus rectum 10.

19. Center $(0,0)$, transverse axis along the y axis, passing through the points $(5,3)$ and $(-3,2)$.

20. Center $(0,0)$, transverse axis along the y axis, distance between the foci is 16, distance between the vertices is 12.

21. Center $(1,-2)$, transverse axis parallel to the x axis, transverse axis 6, conjugate axis 10.

22. Center $(-3,2)$, transverse axis parallel to the y axis, passing through $(1,7)$, the asymptotes are perpendicular to each other.

23. Center $(0,6)$, conjugate axis along the y axis, asymptotes are $6x - 5y + 30 = 0$ and $6x + 5y - 30 = 0$.

24. Derive Eq. (13), (14), and (15), Section 8.

25. The pair of hyperbolas

$$\frac{x^2}{a^2} - \frac{y^2}{b^2} = 1 \qquad \text{and} \qquad \frac{y^2}{b^2} - \frac{x^2}{a^2} = 1$$

are called **conjugate hyperbolas.** Show that conjugate hyperbolas have the same asymptotes. Are conjugate hyperbolas shaped alike when $a = b$? Sketch, on the same coordinate axes, the pair of conjugate hyperbolas corresponding to $a = 5$ and $b = 3$.

26. Show that $(c, 0)$ is a focus and $x = \dfrac{a^2}{c}$ is a directrix of the hyperbola

$$\frac{x^2}{a^2} - \frac{y^2}{b^2} = 1.$$

[Hint: Proceed as in the case of the ellipse (Section 6)].

27. Listening posts are at A, B, and C. Point A is 2,000 ft north of point B, and point C is 2,000 ft east of B. The sound of a gun reaches A and B simultaneously one second after it reaches C. Show that the coordinates of the gun's position are approximately $(860, 1000)$, where the x axis passes through B and C and the origin is midway between B and C. Assume that sound travels 1,100 ft/sec.

28. Find the equation of the locus of a point $P(x, y)$ which moves so that its distance from $(5, 0)$ is equal to $5/4$ its distance from the line $x = 16/5$.

29. Find the equation of the graph of a point which moves so that its distance from $(4, 0)$ is twice its distance from the line $x = 1$.

30. If a and b are constants and p is a parameter such that $p > 0$ and $b^2 - p > 0$, show that the equation

$$\frac{x^2}{a^2 + p} - \frac{y^2}{b^2 - p} = 1$$

represents a family of hyperbolas with common foci on the x axis.

31. Write an equation of the family of hyperbolas whose foci are $(-4, 0)$ and $(4, 0)$. Find the member of the family which passes through $(2, 0)$.

Chapter 4

Simplification of Equations

1. SIMPLIFICATION BY TRANSLATION

In Chapter 2, Section 8 we discovered that the equation of a circle whose center is not at the origin can be expressed in a simpler form by an appropriate translation of axes. Also, in Chapter 3 we used the idea of translation of axes to express the equations of parabolas, ellipses, and hyperbolas in standard forms. We might surmise, then, that general equations of conics are expressible in simple forms. This is indeed true, as we shall see.

The general equation of the second degree may be presented in the form

$$\boxed{Ax^2 + Bxy + Cy^2 + Dx + Ey + F = 0} \tag{1}$$

But before taking up the general equation we shall deal with special cases obtained by setting one or more of the coefficients equal to zero. We advise the student to review the derivation of the translation formulas (Chapter 2, Section 8) and then to study carefully the following examples.

Example 1. By a translation of axes, simplify the equation

$$x^2 - 6x - 6y - 15 = 0$$

Solution. We complete the square in the x terms and select the translation which will eliminate the x' term and the constant term in the new equation. Thus, we get

$$x^2 - 6x + 9 = 6y + 15 + 9$$

$$(x - 3)^2 = 6(y + 4)$$

The vertex of the parabola is at $(3, -4)$. So we put $h = 3$ and $k = -4$ in the translation formulas $x = x' + h$ and $y = y' + k$, and have

$$(x' + 3 - 3)^2 = 6(y' - 4 + 4)$$

$$x'^2 = 6y'$$

The parabola in Fig. 1 is the graph of the given equation referred to the original axes and also the graph of the new equation referred to the new axes.

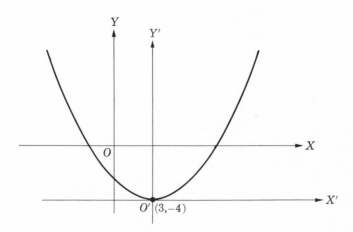

Figure 1

Example 2. Translate the axes so as to simplify the equation

$$2x^2 + 3y^2 + 10x - 18y + 26 = 0$$

Solution. The simplification can be made by completing the squares in the x and y terms. By this plan, we obtain

$$2(x^2 + 5x) + 3(y^2 - 6y) = -26$$

$$2(x^2 + 5x + \tfrac{25}{4}) + 3(y^2 - 6y + 9) = 26 + \tfrac{25}{2} + 27$$

$$2(x + \tfrac{5}{2})^2 + 3(y - 3)^2 = \tfrac{27}{2}$$

The translation formulas $x = x' - \tfrac{5}{2}$ and $y = y' + 3$ will reduce this equation to one free of first-degree terms. Making these substitutions, we find

$$2(x' - \tfrac{5}{2} + \tfrac{5}{2})^2 + 3(y' + 3 - 3)^2 = \tfrac{27}{2}$$

$$4x'^2 + 6y'^2 = 27$$

The graph is an ellipse with $a = \frac{3}{2}\sqrt{3}$, $b = \frac{3}{2}\sqrt{2}$, and $c = \frac{3}{2}$. Both sets of axes and the graph are in Fig. 2.

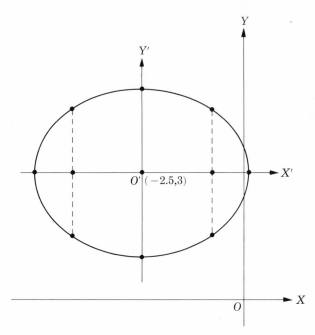

Figure 2

Examples 1 and 2 illustrate the fact that the axes can be translated so as to reduce a second-degree equation of the form

$$Ax^2 + Cy^2 + Dx + Ey + F = 0$$

to one of the forms

$$A'x'^2 + E'y' = 0$$

$$C'y'^2 + D'x' = 0$$

$$A'x'^2 + C'y'^2 + F' = 0$$

In the two examples we completed squares to locate the origins of the new coordinate systems. That plan would need to be altered when an xy term is present.

Example 3. Translate the axes to rid the equation $2xy + 3x - 4y = 12$ of first-degree terms.

Solution. For this equation we use the translation formulas with h and k unknown. The equation then becomes

$$2(x' + h)(y' + k) + 3(x' + h) - 4(y' + k) = 12$$

or, by multiplying and collecting terms,

$$2x'y' + (2k + 3)x' + (2h - 4)y' + 2hk + 3h - 4k = 12$$

To eliminate the x' term and the y' term, we equate their coefficients to zero, thus obtaining $h = 2$ and $k = -\frac{3}{2}$. Using these values for h and k, we get

$$x'y' = 3$$

We shall learn in Section 4 that the graph of this equation, and the given equation, is a hyperbola. Both sets of axes and the hyperbola are drawn in Fig. 3.

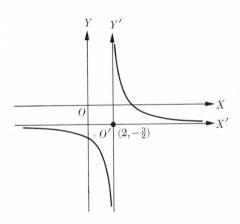

Figure 3

EXERCISES

Determine the new equation in each of Exercises 1 through 12 if the origin is translated to the given point.

1. $3x + 2y = 6$, $(4, -3)$

   $3\left(x' + 4\right) + 2\left(y' - 3\right) = 6$

   $3x' + 2y' = 0$

2. $5x - 4y + 3 = 0$, $(1, 2)$

3. $y^2 - 6x - 4y + 22 = 0$, $(3, 2)$

4. $x^2 - 4x - 7y + 46 = 0$, $(2, 6)$

5. $3x^2 + 4y^2 + 12x + 8y + 8 = 0$, $(-2,-1)$

6. $9x^2 + y^2 + 36x - 8y + 43 = 0$, $(-2,4)$

7. $4y^2 - 5x^2 - 8y - 10x - 21 = 0$, $(-1,1)$

8. $16x^2 - 4y^2 - 160x - 24y + 300 = 0$, $(5,-3)$

9. $xy - x - y - 10 = 0$, $(1,1)$

10. $3xy - 21x + 6y - 47 = 0$, $(-2,7)$

11. $x^3 - 3x^2 + 3x - y - 3 = 0$, $(1,-2)$

12. $3x^3 - 18x^2 + 36x - 4y - 36 = 0$, $(2,-3)$

In each of Exercises 13 through 22, find the point to which the origin must be translated in order that the transformed equation shall have no first-degree term. Find also the new equation.

13. $xy - 2x - 4y - 4 = 0$

14. $xy + 3x - 3y - 3 = 0$

15. $2x^2 + 2y^2 - 8x + 5 = 0$

16. $x^2 + 2y^2 + 6x + 4y + 2 = 0$

17. $3x^2 - 2y^2 + 24x - 8y - 34 = 0$

18. $2y^2 - 3x^2 - 12x + 16y + 14 = 0$

19. $x^2 - xy + y^2 - 9x + 6y - 27 = 0$

20. $2x^2 - 3xy - y^2 - x + 5y - 3 = 0$

21. $x^3 + 5x^2 + 2xy + 4x + 4y - 4 = 0$

22. $x^3 - 6x^2 + xy + 12x - 2y - 7 = 0$

In each of Exercises 23 through 28, eliminate the constant term and one of the first-degree terms.

23. $y^2 - 6y + 4x + 5 = 0$

24. $x^2 - 2x - 8y - 15 = 0$

25. $y^2 + 10x + 4y + 24 = 0$

26. $y^2 - 4y - x + 1 = 0$

27. $2x^2 - 20x - 7y + 36 = 0$

28. $3y^2 + 11x + 6y - 19 = 0$

2. ROTATION OF AXES

We now wish to consider a transformation of coordinates where the new axes have the same origin but different directions from the original axes. Since the new axes may be obtained by rotating the original axes through an angle about the origin, the transformation is called a **rotation of axes**.

We shall derive transformation formulas, for a rotation through an angle θ, which express the old coordinates in terms of θ and the new coordinates. In Fig. 4 the coordinates of the point P are (x,y) referred to the original axes OX and OY, and are (x',y') when referred to the new axes OX' and OY'. We

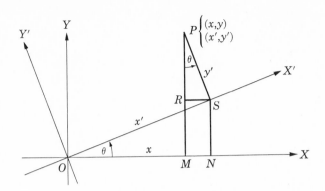

Figure 4

notice that $x = \overrightarrow{OM}$ and $y = \overrightarrow{MP}$, $x' = \overrightarrow{OS}$ and $y' = \overrightarrow{SP}$. The segment $\overrightarrow{RS}$ is drawn parallel to the x axis and $\overrightarrow{NS}$ is parallel to the y axis. Hence we have

$$x = \overrightarrow{OM} = \overrightarrow{ON} - \overrightarrow{MN} = \overrightarrow{ON} - \overrightarrow{RS} = x'\cos\theta - y'\sin\theta$$

$$y = \overrightarrow{MP} = \overrightarrow{MR} + \overrightarrow{RP} = \overrightarrow{NS} + \overrightarrow{RP} = x'\sin\theta + y'\cos\theta$$

The rotation formulas, therefore, are

$$\boxed{\begin{aligned} x &= x'\cos\theta - y'\sin\theta \\ y &= x'\sin\theta + y'\cos\theta \end{aligned}} \tag{2}$$

We have derived these formulas for the special case in which θ is an acute angle and the point P is in the first quadrant of both sets of axes. The formulas hold, however, for any θ and for all positions of P. A proof that the formulas hold generally could be made by observing the proper conventions as to the sign of θ and the signs of all distances involved.

Example 1. Transform the equation $x^2 - y^2 - 9 = 0$ by rotating the axes through $45°$.

Solution. When $\theta = 45°$, the rotation formulas (2) are

$$x = \frac{x'}{\sqrt{2}} - \frac{y'}{\sqrt{2}} \qquad y = \frac{x'}{\sqrt{2}} + \frac{y'}{\sqrt{2}}$$

We make the substitutions in the given equation and have

$$\left(\frac{x'}{\sqrt{2}} - \frac{y'}{\sqrt{2}}\right)^2 - \left(\frac{x'}{\sqrt{2}} + \frac{y'}{\sqrt{2}}\right)^2 - 9 = 0$$

$$\frac{x'^2}{2} - x'y' + \frac{y'^2}{2} - \frac{x'^2}{2} - x'y' - \frac{y'^2}{2} - 9 = 0$$

$$2x'y' + 9 = 0$$

The graph and both sets of axes are constructed in Fig. 5.

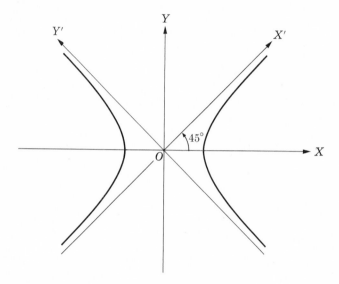

Figure 5

Example 2. Find the acute angle of rotation such that the transformed equation of $2x^2 + \sqrt{3}xy + y^2 = 8$ will have no $x'y'$ term.

Solution. We employ the rotation formulas to find the required angle θ. Substituting for x and y, we get

$$2(x'\cos\theta - y'\sin\theta)^2 + \sqrt{3}(x'\cos\theta - y'\sin\theta)(x'\sin\theta + y'\cos\theta)$$

$$+ (x'\sin\theta + y'\cos\theta)^2 = 8$$

We perform the indicated multiplications, collect like terms, and obtain

$$(2\cos^2\theta + \sqrt{3}\sin\theta\cos\theta + \sin^2\theta)x'^2 + (-2\sin\theta\cos\theta + \sqrt{3}\cos^2\theta$$

$$- \sqrt{3}\sin^2\theta)x'y' + (2\sin^2\theta - \sqrt{3}\sin\theta\cos\theta + \cos^2\theta)y'^2 = 8 \quad (3)$$

Since the $x'y'$ term is to vanish, we set its coefficient equal to zero. Thus we have

$$-2 \sin \theta \cos \theta + \sqrt{3}(\cos^2 \theta - \sin^2 \theta) = 0$$

Using the identities $\sin 2\theta = 2 \sin \theta \cos \theta$ and $\cos 2\theta = \cos^2 \theta - \sin^2 \theta$, we obtain the equation in the form

$$-\sin 2\theta + \sqrt{3} \cos 2\theta = 0$$

Hence

$$\tan 2\theta = \sqrt{3} \qquad 2\theta = 60° \qquad \theta = 30°$$

A rotation of 30° eliminates the $x'y'$ term. This value of θ reduces Eq. (3) to

$$5x'^2 + y'^2 = 16$$

Figure 6 shows both sets of axes and the graph.

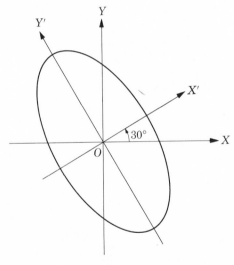

Figure 6

EXERCISES

Find the new equation in Exercises 1 through 8 when the axes are rotated through the given angle. Sketch the graph of the new equation, showing both sets of axes.

1. $\sqrt{3}x - y = 4$, $\theta = 60°$

2. $x + y = 6$, $\theta = 45°$

3. $xy = 4$, $\theta = 45°$

4. $x^2 + y^2 = a^2$, $\theta = 40°$

5. $x^2 + xy + y^2 = 1,\ \theta = 45°$ 6. $x^2 - \sqrt{3}xy + 2y^2 = 2,\ \theta = 30°$

7. $x^2 - 4xy + 4y^2 - 8\sqrt{5}x - 4\sqrt{5}y = 0,\ \theta = \arctan \frac{1}{2}$

8. $x^2 + \sqrt{3}xy + 2y^2 = 3,\ \theta = \arctan \sqrt{3}$

Find the angle of rotation in each of Exercises 9 through 12 such that the transformed equation will have no $x'y'$ term.

9. $3xy + y - 2 = 0$ 10. $x^2 - xy + 5 = 0$

11. $x^2 - 3xy + 4y^2 + 7 = 0$ 12. $x^2 + 3xy - x + y = 0$

3. SIMPLIFICATIONS BY ROTATIONS AND TRANSLATIONS

The general second-degree, or quadratic, equation in x and y is represented by

$$Ax^2 + Bxy + Cy^2 + Dx + Ey + F = 0 \qquad (1)$$

At least one of the constants A, B, and C must be different from zero for the equation to be of the second degree. We assume, too, that not all the coefficients of terms involving one of the variables are zero; that is, both x and y appear in the equation.

In Section 1 we discussed the procedure for simplifying Eq. (1) when $B = 0$. If $B \neq 0$, an essential part of the simplification consists in obtaining a transformed equation lacking the product term $x'y'$. We shall show how to determine immediately an angle of rotation which will serve for this purpose. In Eq. (1) we substitute the right members of the rotation formulas for x and y. After collecting like terms, we obtain

$$A'x'^2 + B'x'y' + C'y'^2 + D'x' + E'y' + F' = 0$$

where the new coefficients are

$$A' = A \cos^2 \theta + B \sin \theta \cos \theta + C \sin^2 \theta$$
$$B' = B \cos 2\theta - (A - C) \sin 2\theta$$
$$C' = A \sin^2 \theta - B \sin \theta \cos \theta + C \cos^2 \theta$$
$$D' = D \cos \theta + E \sin \theta$$
$$E' = E \cos \theta - D \sin \theta$$
$$F' = F$$

The $x'y'$ term will vanish only if its coefficient is zero. Hence θ must satisfy the equation $B' = B \cos 2\theta - (A - C) \sin 2\theta = 0$. If $A \neq C$, the solution is

$$\tan 2\theta = \frac{B}{A - C}$$

This formula yields the angle of rotation except when $A = C$. If $A = C$, the coefficient of $x'y'$ is $B\cos 2\theta$. Then the term vanishes by giving θ the value 45°. Thus we see that an equation of the form (1) with an xy term can be transformed into an equation free of the product term $x'y'$.

We summarize the preceding results in the following theorem.

THEOREM 1   *A second-degree equation*

$$Ax^2 + Bxy + Cy^2 + Dx + Ey + F = 0$$

in which $B = 0$ can be transformed by a translation of axes into one of the forms

$$
\begin{aligned}
A'x'^2 + C'y'^2 + F &= 0 \\
A'x'^2 + E'y' &= 0 \\
C'y'^2 + D'x' &= 0
\end{aligned}
\tag{4}
$$

If $B \neq 0$, one of these forms can be obtained by a rotation and a translation (if necessary). The angle of rotation θ (chosen acute) is obtained from the equation

$$\tan 2\theta = \frac{B}{A - C}, \quad \text{if} \quad A \neq C$$

or

$$\theta = 45°, \quad \text{if} \quad A = C$$

By this theorem we see how to find the value of $\tan 2\theta$. The rotation formulas, however, contain $\sin \theta$ and $\cos \theta$. These functions can be obtained from the trigonometric identities

$$\sin \theta = \sqrt{\frac{1 - \cos 2\theta}{2}} \qquad \cos \theta = \sqrt{\frac{1 + \cos 2\theta}{2}}$$

The positive sign is selected before each radical because we shall restrict θ to an acute angle.

We can interpret geometrically the transformations which reduce the equation of a conic to one of the simplified forms. The rotation orients the coordinate axes in the directions of the axes of an ellipse or hyperbola, and the translation brings the origin to the center of the conic. For a parabola, the rotation makes one coordinate axis parallel to the axis of the parabola, and the translation moves the origin to the vertex.

Although Eqs. (4) above usually represent conics, there are exceptional cases, depending on the values of the coefficients. The exceptional cases are not our main interest, but we do observe them. The first of the equations has no graph if A', C', and F' all have the same sign, for then there are no real values of x' and y' for which the terms of the left member add to zero. If $F' = 0$ and A' and C' have the same sign, only the coordinates of the origin satisfy the equation. The values for the coefficients may be selected arbitrarily so that the equation represents two intersecting lines, two parallel lines, or one line. Each of the other equations always has a graph, but the graph is a line if the coefficient of the first-degree term is zero. The point and line graphs of second-degree equations are called **degenerate** conics, as was previously noted.

Example 1. Express, in one of the forms (4), the equation

$$3x^2 + 2\sqrt{3}xy + y^2 - 2x - 2\sqrt{3}y - 2 = 0$$

Solution. We first rotate the axes so as to eliminate the $x'y'$ terms. Thus, we have

$$\tan 2\theta = \frac{B}{A - C} = \frac{2\sqrt{3}}{3 - 1} = \sqrt{3}$$

Hence $2\theta = 60°$, $\theta = 30°$, and the formulas for the rotation are

$$x = \frac{\sqrt{3}x'}{2} - \frac{y'}{2} \qquad y = \frac{x'}{2} + \frac{\sqrt{3}y'}{2}$$

Substituting for x and y in the given equation, we obtain

$$3\left(\frac{\sqrt{3}x'}{2} - \frac{y'}{2}\right)^2 + 2\sqrt{3}\left(\frac{\sqrt{3}x'}{2} - \frac{y'}{2}\right)\left(\frac{x'}{2} + \frac{\sqrt{3}y'}{2}\right) + \left(\frac{x'}{2} + \frac{\sqrt{3}y'}{2}\right)^2$$

$$-2\left(\frac{\sqrt{3}x'}{2} - \frac{y'}{2}\right) - 2\sqrt{3}\left(\frac{x'}{2} + \frac{\sqrt{3}y'}{2}\right) - 2 = 0$$

We now pick out the coefficients of each variable stemming from this equation. Ignoring the $x'y'$ terms which add to zero, we find

$$(\tfrac{9}{4} + \tfrac{6}{4} + \tfrac{1}{4})x'^2 + (\tfrac{3}{4} - \tfrac{6}{4} + \tfrac{3}{4})y'^2 + (-\sqrt{3} - \sqrt{3})x' + (1 - 3)y' - 2 = 0$$

and

$$4x'^2 - 2\sqrt{3}x' - 2y' - 2 = 0$$

Dividing by 2 and completing the square in the x' terms, we get

$$2\left(x'^2 - \frac{\sqrt{3}}{2}x' + \tfrac{3}{16}\right) = y' + 1 + \tfrac{3}{8}$$

or

$$2\left(x' - \frac{\sqrt{3}}{4}\right)^2 = y' + \tfrac{11}{8}$$

A translation of axis which moves the origin to the point $\left(\dfrac{\sqrt{3}}{4}, -\dfrac{11}{8}\right)$ yields the final equation

$$2x''^2 = y''$$

The graph is the hyperbola drawn in Fig. 7 along with the three sets of axes.

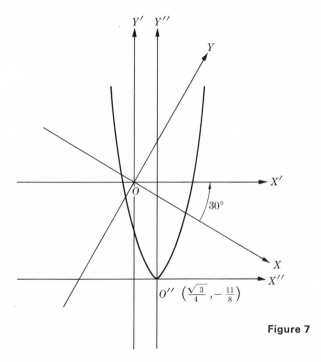

Figure 7

Example 2. Reduce the equation

$$73x^2 - 72xy + 52y^2 + 100x - 200y - 100 = 0$$

to one of the forms (4).

Solution. We first transform the equation so that the product term $x'y'$ will disappear. To find the angle of rotation, we use

$$\tan 2\theta = \frac{B}{A-C} = \frac{-72}{73-52} = \frac{-24}{7}$$

and

$$\cos 2\theta = \frac{-7}{25}$$

Hence

$$\sin \theta = \sqrt{\frac{1 - \cos 2\theta}{2}} = \frac{4}{5} \quad \text{and} \quad \cos \theta = \sqrt{\frac{1 + \cos 2\theta}{2}} = \frac{3}{5}$$

The rotation formulas then become

$$x = \frac{3x' - 4y'}{5} \quad \text{and} \quad y = \frac{4x' + 3y'}{5}$$

By substituting for x and y in the given equation and simplifying, we get

$$x'^2 + 4y'^2 - 4x' - 8y' + 4 = 0$$

Completing the squares in the x' and y' terms, we obtain

$$(x' - 2)^2 + 4(y' - 1)^2 - 4 = 0$$

Finally, a translation to the point $(2, 1)$ yields the desired form

$$x''^2 + 4y''^2 - 4 = 0$$

It is much easier to draw the graph from this equation than by using the original equation. The graph and the three sets of axes are constructed in Fig. 8.

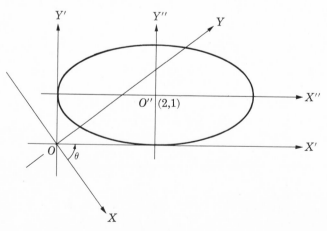

Figure 8

EXERCISES

Reduce each of the equations to one of the simplified forms (4). Draw the graph of the resulting equation, using the last set of coordinate axes. Show all sets of coordinate axes used.

1. $x^2 - 2xy + y^2 - 8\sqrt{2}x - 8 = 0$

2. $x^2 + 4xy + 4y^2 - 2\sqrt{5}x - \sqrt{5}y = 0$

3. $7x^2 + 48xy - 7y^2 - 150x - 50y + 100 = 0$

4. $3x^2 - 10xy + 3y^2 + 22x - 26y + 43 = 0$

5. $30x^2 - 12xy + 35y^2 - 60x + 12y - 48 = 0$

6. $73x^2 - 72xy + 52y^2 + 380x - 160y + 400 = 0$

7. $104x^2 + 60xy + 41y^2 - 60x - 82y - 75 = 0$

8. Prove that a translation of coordinates does not introduce any second-degree terms in the equation of a conic, and that a rotation does not introduce any first-degree terms.

4. IDENTIFICATION OF A CONIC

We have seen that the graph, if any, of the general second-degree equation

$$Ax^2 + Bxy + Cy^2 + Dx + Ey + F = 0 \qquad (1)$$

can be determined immediately when the equation is reduced to one of the simplified forms (4). The determination can also be made from the coefficients of the general equation. To show that this is true, let us first assume that the graph of Eq. (1) is a parabola, an ellipse, or a hyperbola. Then applying the rotation formulas (2), we obtain

$$A'x'^2 + B'x'y' + C'y'^2 + D'x' + E'y' + F' = 0$$

where

$$A' = A\cos^2\theta + B\sin\theta\cos\theta + C\sin^2\theta$$

$$B' = B\cos 2\theta - (A - C)\sin 2\theta$$

$$C' = A\sin^2\theta - B\sin\theta\cos\theta + C\cos^2\theta$$

If the expressions for A', B', and C' are substituted in $B'^2 - 4A'C'$ and simplified, the result is

$$B'^2 - 4A'C' = B^2 - 4AC$$

This relation among the coefficients of the original equation and the transformed equation holds for any rotation. For this reason, $B^2 - 4AC$ is called an **invariant**. By selecting the particular rotation for which $B' = 0$, we have

$$\boxed{-4A'C' = B^2 - 4AC}$$

With $B' = 0$ the kind of conic represented by the transformed equation, and therefore the original equation, can be determined from the signs of A' and C'. The conic is an ellipse if A' and C' have like signs, and a hyperbola if the signs are different. If either A' or C' is zero, the conic is a parabola. These relations of A' and C', in the order named, would make $-4A'C'$ negative, positive, or zero. Hence we have the following important theorem.

THEOREM 2   *Let the coefficients of the equation*

$$Ax^2 + Bxy + Cy^2 + Dx + Ey + F = 0 \tag{1}$$

be such that the graph is a nondegenerate conic. Then the graph is

1. *an ellipse if $B^2 - 4AC < 0$.*
2. *a hyperbola if $B^2 - 4AC > 0$.*
3. *a parabola if $B^2 - 4AC = 0$.*

This theorem is based on the condition that the equation has a graph and that the graph is not a degenerate conic. The degenerate conics, as we have already mentioned, consist of two intersecting lines, two parallel lines, one line, and a single point. So, in order to use the theorem with certainty, we need to know how to detect the exceptional cases. A general discussion of this question is a bit tedious; consequently, we will omit it. We shall, however, illustrate a workable plan when A and C are not both zero, and we shall point out the approach for the simpler situation when A and C are both zero and B is different from zero.

Example 1. Determine the nature of the graph, if any, of the equation

$$2x^2 + 7xy + 3y^2 + x - 7y - 6 = 0$$

Solution. We find that $B^2 - 4AC = 49 - 4(2)(3) = 25 > 0$. This positive result does not tell us that the graph is a hyperbola. To make the determination, we treat the equation as a quadratic in x and apply the quadratic formula. Thus we get

$$2x^2 + (7y + 1)x + (3y^2 - 7y - 6) = 0$$

$$x = \frac{-(7y + 1) \pm \sqrt{(7y + 1)^2 - 8(3y^2 - 7y - 6)}}{4}$$

$$x = \frac{-7y - 1 \pm \sqrt{25y^2 + 70y + 49}}{4}$$

$$= \frac{-7y - 1 \pm (5y + 7)}{4}$$

Hence, $x = \dfrac{-y + 3}{2}$ and $x = -3y - 2$ are the solutions. This means that the given equation may be presented in the factored form

$$(2x + y - 3)(x + 3y + 2) = 0$$

If the coordinates of a point make one of these factors equal to zero, they make the product equal to zero and therefore satisfy the original equation. Hence the graph consists of the two lines whose equations are

$$2x + y - 3 = 0 \quad \text{and} \quad x + 3y + 2 = 0$$

Example 2. Test the equation $y^2 - 4xy + 4x^2 - 2x + y - 12 = 0$.

Solution. Solving the equation for y in terms of x, we find the equations

$$y - 2x - 3 = 0 \quad \text{and} \quad y - 2x + 4 = 0$$

Hence the graph of the given equation consists of two parallel lines.

Example 3. Test the equation $4x^2 - 12xy + 9y^2 + 20x - 30y + 25 = 0$.

Solution. We can make the test by solving for one variable in terms of the other. Scrutinizing the equation, though, we see that the left member is readily factorable. Thus, we find

$$(4x^2 - 12xy + 9y^2) + 10(2x - 3y) + 25 = 0$$
$$(2x - 3y)^2 + 10(2x - 3y) + 25 = 0$$
$$(2x - 3y + 5)(2x - 3y + 5) = 0$$

The graph is the line $2x - 3y + 5 = 0$.

Example 4. Determine the graph, if any, of the equation

$$x^2 + y^2 - 4x + 2y + 6 = 0$$

Solution. The student may verify that this equation yields

$$x = 2 \pm \sqrt{-y^2 - 2y - 2}$$

$$= 2 \pm \sqrt{-[(y+1)^2 + 1]}$$

It is evident that the radicand is negative for all real values of y, and this means that x is imaginary. Consequently the given equation has no graph.

Example 5. Determine if the equation

$$x^2 + 3xy + 3y^2 - x + 1 = 0$$

has a graph.

Solution. Solving the equation for x gives

$$x^2 + (3y - 1)x + 3y^2 + 1 = 0$$

$$x = \frac{1 - 3y \pm \sqrt{(3y-1)^2 - 12y^2 - 4}}{2}$$

$$= \frac{1 - 3y \pm \sqrt{-3y^2 - 6y - 3}}{2}$$

$$= \frac{1 - 3y \pm (y+1)\sqrt{-3}}{2}$$

As we see, x will be imaginary if y has any real value except $y = -1$. For this value of y, the value of x is 2. Hence the graph of the given equation consists of the single point $(2, -1)$.

If the squared terms are missing in a second-degree equation of the form (1), we may divide by the coefficient of xy and express the equation as

$$xy + Dx + Ey + F = 0$$

The procedure for testing in this case is to translate the axes so that the new equation will have no x' or y' term. The transformed equation, as the student may verify, is

$$x'y' + (F - DE) = 0$$

Clearly, the graph is a hyperbola if $F - DE \neq 0$. And it is equally evident that the graph consists of two intersecting lines if $F - DE = 0$.

We have pointed out the five ways in which Eq. (1) will not represent a conic. It is important to understand that the equation represents a conic, provided it does not conform to one of these exceptional cases. The $B^2 - 4AC$ test of Theorem 2 is then applicable.

5. ADDITION OF ORDINATES

The presence of an xy term in the second-degree equation usually makes the construction of the graph much more difficult. A table of pairs of values of the variables can be prepared by assigning values to x or y and solving for the corresponding values of the other. This method, however, is very tedious. Another plan would be to rotate the axes and use the new equation and the new axes to draw the graph. But rotation transformations are not short, and usually the process is complicated by cumbersome radicals in the rotation formulas. A third way to handle the situation is to resort to a method known as the **addition of ordinates**. This method is applicable if y is the sum of two expressions in x, for example, $y = f(x) + g(x)$. First we draw separate graphs of $y_1 = f(x)$ and $y_2 = g(x)$. Then the ordinates of the curves, when added graphically, yield ordinates of the graph of the original equation. The utility of this method depends on the ease with which we obtain the two auxiliary graphs. The following example illustrates the process.

Example 1. Draw the graph of the equation

$$2x^2 - 2xy + y^2 + 8x - 12y + 36 = 0$$

Solution. To express y as the sum of two quantities, we treat the equation as a quadratic in y. Thus we have

$$y^2 + (-2x - 12)\, y + (2x^2 + 8x + 36) = 0$$

and solving for y gives

$$y = \frac{2x + 12 \pm \sqrt{(-2x - 12)^2 - 4(2x^2 + 8x + 36)}}{2}$$

$$= x + 6 \pm \sqrt{4x - x^2}$$

We now draw the graphs of the equations

$$y_1 = x + 6 \qquad \text{and} \qquad y_2 = \pm\sqrt{4x - x^2}$$

The graph of the first equation is a line. By squaring and then completing the square in the x terms, we reduce the second equation to $(x - 2)^2 + y_2^2 = 4$. The graph is a circle of radius 2 and center at $(2,0)$. The line and the circle are drawn in Fig. 9. The point D on the graph of the original equation is obtained by adding the ordinates AB and AC; that is, AC is extended by a length equal to AB. The addition of ordinates for this purpose must be algebraic. Thus MN is negative, and the point Q is found by measuring downward from

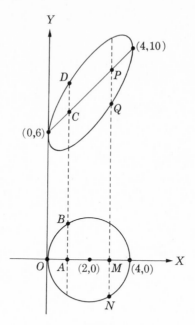

Figure 9

P so that $PQ = MN$. By plotting a sufficient number of points in this manner the desired graph can be constructed.

EXERCISES

Assuming that Eqs. 1 through 8 represent nondegenerate conics, classify each by computing $B^2 - 4AC$.

1. $2x^2 - 4xy + 8y^2 + 7 = 0$

2. $3x^2 + xy + x - 4 = 0$

3. $2xy - x + y - 3 = 0$

4. $x^2 + 5xy + 15y^2 = 1$

5. $x^2 - y^2 + 4 = 0$

6. $x^2 - 2xy + y^2 + 3x = 0$

7. $3x^2 + 6xy + 5y^2 - x + y = 0$

8. $4x^2 - 3xy + y^2 + 13 = 0$

Test each equation in Exercises 9 through 24 and tell whether the graph is a conic, or a degenerate conic, or there is no graph. Give the equation (or equations) of any degenerate graph.

9. $6x^2 - xy - 2y^2 + 2x + y = 0$

10. $y^2 - 8xy + 17x^2 + 4 = 0$

11. $x^2 + 2xy + y^2 - 2x - 2y = 0$

12. $y^2 + 2xy + x^2 - 2x + 1 = 0$

13. $y^2 - 4xy + 4x^2 + 4x + 1 = 2y$

14. $y^2 - 2xy + 2x^2 + 1 = 0$

15. $4y^2 - 12xy + 10x^2 + 2x + 1 = 0$

16. $2x^2 - xy - y^2 + 2x + y = 0$

17. $y^2 - 2xy + 2x^2 - 5x = 0$

18. $9x^2 - 6xy + y^2 + 3x - y = 0$

19. $9x^2 - 6xy + y^2 + x + 1 = 0$

20. $y^2 - 6xy + 11x^2 + 8x + 8 = 0$

21. $xy - 6x - 5y + 30 = 0$

22. $xy + 4x - 4y + 20 = 0$

23. $2x^2 + xy - y^2 + x - y = 0$

24. $3x^2 + 3xy - y^2 - x + y = 0$

Sketch the graph of each equation, Exercises 25 through 30, by the addition of ordinates method.

25. $y = x \pm \sqrt{x}$

26. $y = 6 - x \pm \sqrt{4 - x^2}$

27. $y = 2x \pm \sqrt{5 + 6x - x^2}$

28. $y^2 - 2xy + 2x^2 - 1 = 0$

29. $y^2 - 4xy + 3x^2 + 1 = 0$

30. $y^2 + 2xy - 3x^2 + 4 = 0$

Assume in Exercises 31 through 35 that the graph of

$$Ax^2 + Bxy + Cy^2 + Dx + Ey + F = 0$$

is a nondegenerate conic.

31. If A and C have opposite signs, prove that the graph is a hyperbola.

32. If $B \neq 0$ and either A or C is zero, prove that the graph is a hyperbola.

33. If both D and E are zero, prove that the graph is not a parabola.

34. If both D and E are zero, prove that the center of the conic is at the origin.

35. If the center of the conic is at the origin, prove that both D and E are zero.

36. Work out all steps in showing that $B'^2 - 4A'C' = B^2 - 4AC$. Show also that $A' + C' = A + C$.

Chapter 5

Polar Coordinates

There are various types of coordinate systems. The rectangular system with which we have been dealing is probably the most important. In it a point is located by its distances from two perpendicular lines. We shall introduce in this chapter a coordinate system in which the coordinates of a point in a plane are its distance from a fixed point and its direction from a fixed line. The coordinates given in this way are called **polar coordinates**. The proper choice of a coordinate system depends on the nature of the problem at hand. For some problems either the rectangular or the polar system may be satisfactory; usually, however, one of the two is preferable. And in some situations it is advantageous to use both systems, shifting from one to the other.

1. THE POLAR COORDINATE SYSTEM

The reference frame in the polar coordinate system is a half line drawn from some point in the plane. In Fig. 1 a half line is represented by OA. The point O is called the **origin** or **pole** and OA the **polar axis**. The position of any point P

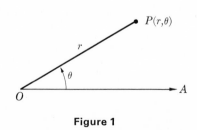

Figure 1

in the plane is definitely determined by the distance $\overrightarrow{OP}$ and the angle AOP. The segment OP, denoted by r, is referred to as the **radius vector**; the angle AOP, denoted by θ, is called the **vectorial angle.** The coordinates of P are then written as $P(r, \theta)$ or just (r, θ).

It is customary to regard polar coordinates as signed quantities. The vectorial angle, as in trigonometry, is defined as positive or negative according as it is measured counterclockwise or clockwise from the polar axis. The r coordinate is defined as positive if measured from the pole along the terminal side of θ and negative if measured along the terminal side extended through the pole.

A given pair of polar coordinates definitely locates a point. For example, the coordinates $(3, 30°)$ determine one particular point. To plot the point, we first draw the terminal side of a $30°$ angle measured counterclockwise from OA (Fig. 2) and then lay off three units along the terminal side. While this pair

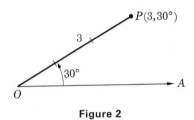

Figure 2

of coordinates defines a particular point, there are other coordinate values which define this same point. This is evident, since the vectorial angle may have $360°$ added or subtracted repeatedly without changing the point represented. Additional coordinates of the point may be obtained also by using a negative value for the distance coordinate. Restricting the vectorial angle to values between $-360°$ and $360°$, we see (Figs. 2 through 5) that the following pairs of coordinates define the same point:

$$(3, 30°) \qquad (3, -330°) \qquad (-3, 210°) \qquad (-3, -150°)$$

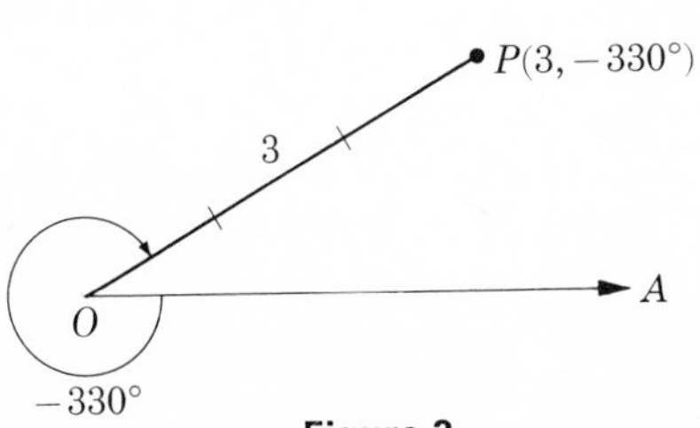

$P(3, -330°)$

3

O

A

$-330°$

Figure 3

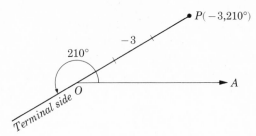

Figure 4

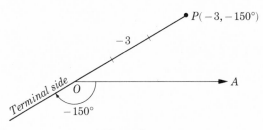

Figure 5

Although we have used here the vectorial angles in terms of degree measure, radian measure would serve just as well. The relation between degrees and radians can be easily determined. The entire circumference of a circle subtends a central angle of 2π in terms of radians and 360 units in terms of degrees. Hence we have

$$2\pi \text{ radians} = 360° \qquad \text{and} \qquad \pi \text{ radians} = 180°$$

We use the second equation to express a radian in terms of degrees and a degree in terms of radian measure. Thus

$$1 \text{ radian} = \frac{180°}{\pi} \qquad \text{and} \qquad 1° = \frac{\pi}{180} \text{ radian}$$

When an angle is expressed in radians, the word "radian" is frequently omitted. As examples, we write

$$\pi = 180° \qquad \frac{\pi}{6} = 30° \qquad \frac{\pi}{4} = 45° \qquad \frac{5\pi}{3} = 300°$$

We also allow ourselves the convenience of using "angle" and "measure of an angle" interchangeably. For example, if θ stands for an angle of measure 60°, we write $\theta = 60°$ when really the measure of the angle is 60°.

A point of given coordinates can be plotted by estimating the vectorial angle and the radius vector by sight. Greatly improved accuracy, however, may be obtained by the use of polar coordinate paper. This paper has equally spaced circles with their centers at the origin and equally spaced radial lines through the origin (Fig. 6). A point of given coordinates may be located by selecting the proper θ and r coordinates. Several points are plotted in the figure.

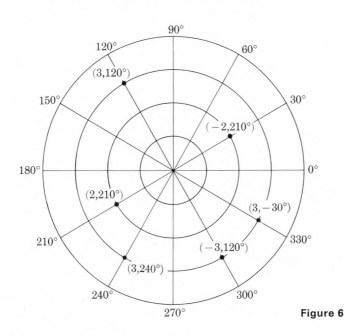

Figure 6

EXERCISES

Express each of the following angles in terms of π radians.

1. $60°, 135°, 150°$ 2. $30°, 90°, 240°$

3. $225°, 315°, 330°$

Change each angle from radian measure to degrees.

4. $\dfrac{3\pi}{2}, \dfrac{\pi}{10}, \dfrac{\pi}{15}$ 5. $\dfrac{\pi}{9}, \dfrac{3\pi}{4}, \dfrac{7\pi}{24}$ 6. $\dfrac{11\pi}{12}, \dfrac{3\pi}{8}, \dfrac{4\pi}{3}$

Plot the given points on a polar coordinate system.

7. a) $P(3, 60°)$, b) $P(6, -30°)$, c) $P(2, 180°)$, d) $P(4, 15°)$

8. a) $P(5, 210°)$, b) $P(4, 0°)$, c) $P(-3, 135°)$, d) $P(-2, -180°)$

9. a) $P(-1, \frac{5}{6}\pi)$, b) $P(-3, \frac{4}{3}\pi)$, c) $P(-4, \pi)$, d) $P(4, -\frac{3}{2}\pi)$

10. a) $P(2, \frac{5}{3}\pi)$, b) $P(-2, \frac{5}{4}\pi)$, c) $P-3, \frac{7}{4}\pi)$, d) $P(3, -\frac{23}{12}\pi)$

11. Write three other pairs of polar coordinates for the points in Exercise 7. Restrict the vectorial angles so that they do not exceed 360° in absolute value.

12. Write three other pairs of polar coordinates for the points in Exercise 9. Restrict the vectorial angles θ so that $-2\pi \leqq \theta \leqq 2\pi$.

2. RELATIONS BETWEEN RECTANGULAR AND POLAR COORDINATES

As we have mentioned, it is often advantageous in the course of a problem to shift from one coordinate system to another. For this purpose we shall derive transformation formulas which express polar coordinates in terms of rectangular coordinates, and vice versa. In Fig. 7 the two systems are placed so that the

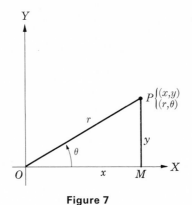

Figure 7

origins coincide and the polar axis lies along the positive x axis. Then a point P has the coordinates (x, y) and (r, θ). From the triangle OMP, we have

$$\cos \theta = \frac{x}{r} \quad \text{and} \quad \sin \theta = \frac{y}{r}$$

and hence

$$x = r \cos \theta \tag{1}$$

$$y = r \sin \theta \tag{2}$$

To obtain r and θ in terms of x and y, we write

$$r^2 = x^2 + y^2 \qquad \text{and} \qquad \tan\theta = \frac{y}{x}$$

Then, solving for r and θ, we get

$$\boxed{r = \pm\sqrt{x^2 + y^2}} \tag{3}$$

$$\boxed{\theta = \arctan\frac{y}{x}} \tag{4}$$

These four equations enable us to transform the coordinates of a point, and therefore the equation of a graph, from one system to the other. The θ coordinate as given by Eq. (4) is not single-valued. Hence it is necessary to select a proper value for θ when applying the formula to find this coordinate of a point. See Example 2.

Example 1. Find the rectangular coordinates of the point defined by the polar coordinates $(6, 120°)$.

Solution. Using Eqs. (1) and (2), we have (Fig. 8)

$$x = r\cos\theta = 6\cos 120° = -3$$

$$y = r\sin\theta = 6\sin 120° = 3\sqrt{3}$$

The required coordinates are $(-3, 3\sqrt{3})$.

Example 2. Express the rectangular coordinates $(-2, -2)$ in terms of polar coordinates

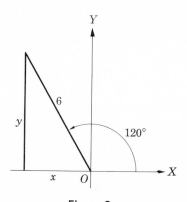

Figure 8

Solution. Equations (3) and (4) give

$$r = \sqrt{x^2 + y^2} = 2\sqrt{2} \qquad \text{and} \qquad \theta = \arctan\frac{y}{x} = \arctan 1$$

Since the point is in the third quadrant, we select $\theta = 225°$. Hence the pair of coordinates $(2\sqrt{2}, 225°)$ is a polar representation of the given point.

Example 3. Find the polar coordinate equation corresponding to $2x - 3y = 5$.

Solution. Substituting for x and y gives

$$2(r\cos\theta) - 3(r\sin\theta) = 5 \qquad \text{or} \qquad r(2\cos\theta - 3\sin\theta) = 5$$

Example 4. Transform the equation $r = 4\sin\theta$ to rectangular coordinates.

Solution. Since $r = \sqrt{x^2 + y^2}$ and $\sin\theta = y/r = y/\sqrt{x^2 + y^2}$, we substitute in the given equation and obtain

$$\sqrt{x^2 + y^2} = \frac{4y}{\sqrt{x^2 + y^2}}$$

or

$$x^2 + y^2 = 4y$$

The required equation, as well as the original equation, represents a circle.

Example 5. Transform the polar coordinate equation

$$r = \frac{1}{\cos\theta + 3\sin\theta}$$

to the corresponding rectangular coordinate equation.

Solution. We multiply both members of the equation by $\cos\theta + 3\sin\theta$ and obtain

$$r(\cos\theta + 3\sin\theta) = 1$$

or, from Eqs. (1) and (2)

$$r\left(\frac{x}{r} + \frac{3y}{r}\right) = 1$$

Now, multiplying, we have the linear equation

$$x + 3y = 1$$

The final equation tells us that the graph of the given equation is a straight line

EXERCISES

1. Derive the transformation formulas of Eqs. (1) through (4) from a figure in which the terminal side is in the (a) second quadrant, (b) third quadrant, (c) fourth quadrant.

Find the rectangular coordinates of the following points.

2. $(4, 90°)$ 3. $(3\sqrt{2}, 45°)$ 4. $(7, 0°)$

5. $(0, 180°)$ 6. $(-8, 270°)$ 7. $(-1, -60°)$

8. $(6, 150°)$ 9. $(4\sqrt{2}, -135°)$ 10. $(9, 180°)$

Find nonnegative polar coordinates of the following points.

11. $(0, 3)$ 12. $(3, 0)$ 13. $(0, 0)$

14. $(-1, 0)$ 15. $(0, -5)$ 16. $(\sqrt{2}, \sqrt{2})$

17. $(6\sqrt{3}, -6)$ 18. $(-2\sqrt{3}, 2)$ 19. $(3, -4)$

20. $(-4, 3)$ 21. $(5, 12)$ 22. $(-5, -12)$

Transform the following equations to the corresponding polar coordinate equations.

23. $x = 3$ 24. $y = -4$ 25. $2x - y = 3$

26. $3x + y = 0$ 27. $y^2 = 4x$ 28. $2xy = a^2$

29. $x^2 + y^2 = 16$ 30. $x^2 - y^2 = a^2$ 31. $x^2 - 9y = 0$

32. $x^2 + 2y^2 = 6$ 33. $x^2 + y^2 = 2x$ 34. $(x^2 + y^2)^2 = 2xy$

Transform the equations to the corresponding rectangular coordinate equations.

35. $r = 4$ 36. $\theta = 45°$ 37. $\theta = 60°$

38. $r \cos\theta = 6$ 39. $r \sin\theta = 6$ 40. $r = 8 \cos\theta$

41. $r = 8 \sin\theta$ 42. $r^2 \cos 2\theta = a^2$ 43. $r^2 \sin 2\theta = a^2$

44. $r = \dfrac{3}{1 + \cos\theta}$ 45. $r = \dfrac{3}{2 + \cos\theta}$ 46. $r = \dfrac{2}{1 - 2\cos\theta}$

47. $r = \dfrac{2}{3\sin\theta + 4\cos\theta}$ 47. $r = \dfrac{5}{2\sin\theta - \cos\theta}$

49. $r = \dfrac{1}{\cos\theta - 3\sin\theta}$ 50. $r = \dfrac{2}{1 - 2\sin\theta}$

3. GRAPHS OF POLAR COORDINATE EQUATIONS

In this section we shall consider the problem of plotting the graphs of equations expressed in polar coordinates. The definition of a graph in polar coordinates and the technique of its construction are essentially the same as with an equation in rectangular coordinates.

DEFINITION 1   *The **graph** of an equation in polar coordinates is the set of all points whose coordinates satisfy the equation.*

The graphs of the equations most easily determined are those in which each coordinate is equal to a constant. For example, the graph of $\theta = 35°$ is the line through the origin and making an angle of $35°$ with the polar axis. If $35°$ is used for the θ coordinate at all points of the line, the corresponding r coordinates will be positive on the terminal side of $35°$ and negative on the extension of the terminal side through the origin. This is illustrated by the points $(4, 35°)$ and $(-4, 35°)$. Similarly, the graph of $r = a$, where a is any real nonzero number, is a circle with center at the origin and radius $|a|$.

Example 1. Construct the graph of $r = 3 + \sin \theta$.

Solution. We assign certain values to θ from $0°$ to $360°$ and prepare a table of corresponding values of θ and r. The fractional values of r are rounded off to one decimal place.

θ	$0°$	$30°$	$45°$	$60°$	$90°$	$120°$	$150°$	$180°$
r	3	3.5	3.7	3.9	4	3.9	3.5	3

$210°$	$225°$	$240°$	$270°$	$300°$	$315°$	$330°$	$360°$
2.5	2.3	2.1	2	2.1	2.3	2.5	3

By plotting these points and drawing a curve through them, the graph of Fig. 9 is obtained.

Example 2. Draw the graph of $r = 4 \cos \theta$.

Solution. We first prepare a table of corresponding values of r and θ.

θ	$0°$	$30°$	$45°$	$60°$	$75°$	$90°$	$120°$	$135°$	$150°$	$180°$
r	4	3.5	2.8	2.0	1.0	0	-2.0	-2.8	-3.5	-4

This table yields the graph in Fig. 10. We did not extend the table to include values of θ in the interval $180°$ to $360°$, since values of θ in this range would merely repeat the graph already obtained. For example, the point $(-3.5, 210°)$ is on the graph, but this point is also defined by the coordinates $(3.5, 30°)$.

The graph appears to be a circle. This surmise is verified by transform-

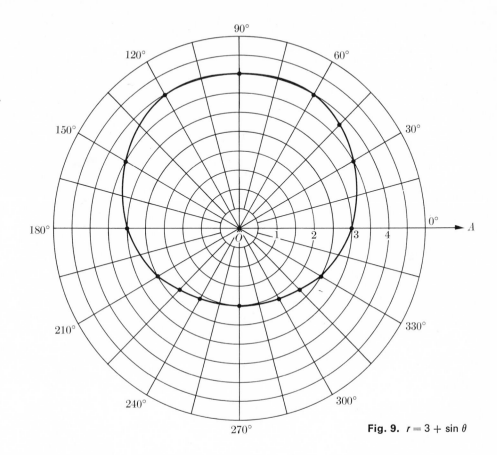

Fig. 9. $r = 3 + \sin \theta$

ing the equation to rectangular coordinates. The transformed equation is
$(x - 2)^2 + y^2 = 4$.

Example 3. Draw the graph of the equation

$$r = \frac{1}{1 + \sin \theta}$$

Solution. The table of corresponding values of r and θ enables us to draw
the curve in Fig. 11.

θ	0°	90°	180°	210°	225°	315°	330°
r	1	0.50	1	2	3.41	3.41	2

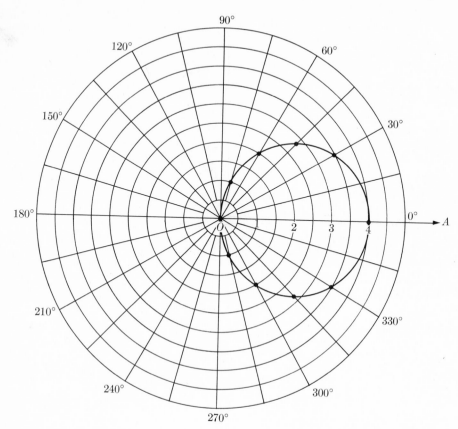

Fig. 10. $r = 4\cos\theta$

EXERCISES

Draw the graphs of the following equations. In the equations involving $\sin\theta$ and $\cos\theta$, points plotted at $30°$ intervals will suffice, with some exceptions.

1. $r = 5$

2. $r = -5$

3. $\theta = 120°$

4. $r = 2 + \cos\theta$

5. $r = 2 - \sin\theta$

6. $r = 3\sin\theta$

7. $r = 1 - \cos\theta$

8. $r = 1 - \sin\theta$

9. $r = -2\cos\theta$

10. $r = \dfrac{2}{2 + \sin\theta}$

11. $r = \dfrac{1}{1 + \cos\theta}$

12. $r = \dfrac{1}{1 - \sin\theta}$

4. AIDS IN GRAPHING POLAR COORDINATE EQUATIONS

A careful examination of an equation in polar coordinates will often reveal shortcuts to the construction of its graph. It is better, for economy of time, to

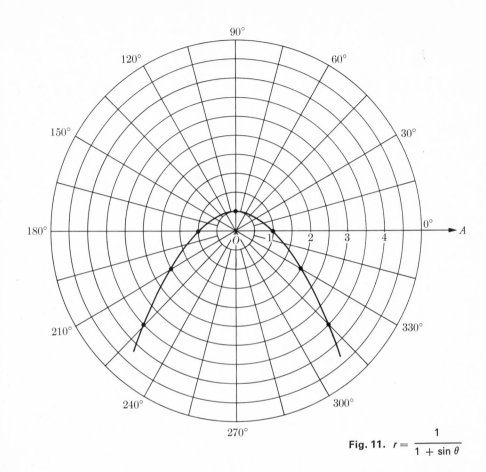

Fig. 11. $r = \dfrac{1}{1 + \sin\theta}$

wrest useful information from an equation and thus keep at a minimum the point-by-point plotting in sketching a graph. We shall discuss and illustrate a few simple devices which facilitate the tracing of polar curves.

Variation of r with θ

Many equations are sufficiently simple so that the way in which r varies as θ increases is evident. Usually the variation of θ from $0°$ to $360°$ yields the complete graph. However, we shall find exceptions to this rule. By observing the equation and letting θ increase through the necessary values, the graph can be visualized. A rough sketch can then be made with a few pencil strokes. We illustrate with an example.

Example 1. Sketch the graph of the equation

$$r = 3(1 + \sin\theta)$$

Solution. If θ starts at $0°$ and increases in $90°$-steps to $360°$, it is a simple matter to see how r varies in each $90°$ interval. This variation is represented in the diagram. The graph (Fig. 12) is a heart-shaped curve called the **cardioid**.

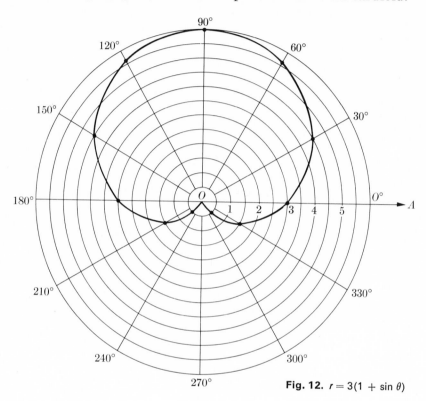

Fig. 12. $r = 3(1 + \sin \theta)$

As θ increases from	$\sin \theta$ varies from	r varies from
$0°$ to $90°$	0 to 1	3 to 6
$90°$ to $180°$	1 to 0	6 to 3
$180°$ to $270°$	0 to -1	3 to 0
$270°$ to $360°$	-1 to 0	0 to 3

Tangent lines at the origin

If r shrinks to zero as θ approaches and takes a fixed value θ_0, then the line $\theta = \theta_0$ is tangent to the curve at the origin. Intuitively, this statement seems correct; it can be proved. In the preceding equation, $r = 3(1 + \sin \theta)$, the value of r diminishes to zero as θ increases to $270°$. Hence the curve is tangent to the vertical line at the origin (Fig. 12).

To find the tangents to a curve at the origin, set $r = 0$ in the equation and solve for the corresponding values of θ.

Symmetry

In Chapter 3, Section 3 we discussed symmetry of curves and formulated tests applicable in a rectangular coordinate system. The following tests for symmetry in a polar coordinate system are deducible by referring to Fig. 13.

1. *If the equation is unchanged when r is replaced by $-r$ or when θ is replaced by $180° + \theta$, the graph is symmetric with respect to the pole.*

2. *If the equation is unchanged when θ is replaced by $-\theta$, the graph is symmetric with respect to the polar axis.*

3. *If the equation is unchanged when θ is replaced by $180° - \theta$, the graph is symmetric with respect to the vertical line $\theta = 90°$.*

These tests will be found helpful. When any of the tests is satisfied in an equation, the symmetry is certain. On the other hand, the failure of a

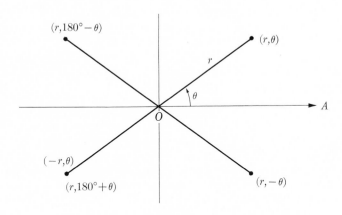

Figure 13

test does not disprove the symmetry in question. This is unlike the analogous situation in rectangular coordinates and is a consequence of the fact that a point has more than one polar coordinate representation. When a graph possesses any two of the three types of symmetry, the remaining type exists. The student should verify this statement by examining Fig. 13.

For the convenience of the student, we list the following trigonometric identities:

$$\sin(-\theta) = -\sin\theta \qquad \cos(-\theta) = \cos\theta$$
$$\sin(180° - \theta) = \sin\theta \qquad \cos(180° - \theta) = -\cos\theta$$
$$\sin(180° + \theta) = -\sin\theta \qquad \cos(180° + \theta) = -\cos\theta$$

Example 2. Construct the graph of $r = \sin 2\theta$.

Solution. We first apply the three tests for symmetry. Replacing r by $-r$ the equation

$$r = \sin 2\theta \qquad \text{becomes} \qquad -r = \sin 2\theta$$

This does not establish symmetry with respect to the pole. But substituting $180° + \theta$ for θ yields

$$r = \sin 2(180° + \theta) = \sin(360° + 2\theta) = \sin 2\theta$$

This result shows that the graph is symmetric with respect to the pole. Applying tests 2 and 3, we obtain

$$r = \sin(-2\theta) = -\sin 2\theta$$

$$r = \sin 2(180° - \theta) = \sin(360° - 2\theta) = -\sin 2\theta$$

Hence these tests fail.

Since there is symmetry with respect to the pole, it is sufficient to obtain the graph for θ from $0°$ to $180°$, and then use the known symmetry to complete the drawings. To determine how r varies, we let θ increase in steps of $45°$. The diagram indicates the variation.

θ	2θ	$\sin 2\theta$, or r
$0° \to 45°$	$0° \to 90°$	$0 \to 1$
$45° \to 90°$	$90° \to 180°$	$1 \to 0$
$90° \to 135°$	$180° \to 270°$	$0 \to -1$
$135° \to 180°$	$270° \to 360°$	$-1 \to 0$

The values of θ corresponding to $r = 0$ are $90°$ and $180°$. This means the graph is tangent to the polar axis and the vertical line at the origin. The completed graph (Fig. 14) exhibits all three types of symmetry. Because of its shape, the

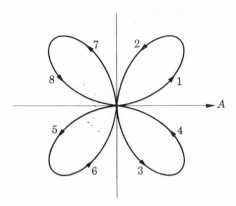

Fig. 14. $r = \sin 2\theta$

graph is called a **four-leaved rose**. The numbered barbs indicate how a point would move in tracing the curve as θ increases from $0°$ to $360°$.

Excluded Values

Frequently we shall meet equations in which certain values of the variables are excluded. For example, $r^2 = a^2 \sin \theta$ places restrictions on both r and θ. The values of r are in the range $-a$ to a, and θ cannot have a value which makes $\sin \theta$ negative since r would then be imaginary. In particular, the angles between $180°$ and $360°$ are excluded. The graph, however, extends into the third and fourth quadrants for other values of θ because the equation satisfies the test for symmetry with respect to the origin.

Intercepts

The points at which a curve touches or cuts the horizontal and vertical lines through the pole are called **intercept** points; they are sometimes quite helpful in drawing the graph of a polar equation. The intercept points can be found by using the values $0°$, $90°$, $180°$, $270°$, or angles coterminal with these, for θ and finding the corresponding values of r. In this way, for example, we find that the intercept points of $r = 3 + \sin \theta$ are $(3, 0°)$, $(4, 90°)$, $(3, 180°)$, and $(2, 270°)$. The intercept points of $r = 2\theta$ are limitless in number. They occur at $\theta = n\pi$ and $\theta = (n + \frac{1}{2})\pi$, where n is any integer, positive, negative, or zero.

We remark, however, that a curve may pass through the origin at angles other than a quadrantal angle. The test for this situation, as already stated, is to find the values of θ when $r = 0$.

5. SPECIAL TYPES OF EQUATIONS

There are several types of polar coordinate equations whose graphs have been given special names. We consider a few of these equations.

The graphs of equations of the forms

$$r = a \sin n\theta \quad \text{and} \quad r = a \cos n\theta$$

where n is a positive integer, greater than 1, are called **rose curves**. The graph of a rose curve consists of equally spaced closed loops extending from the origin. The number of loops, or leaves, depends on the integer n. If n is odd, there are n leaves; if n is even, there are $2n$ leaves. Refer again to Fig. 14.

The graph of an equation of the form

$$r = b + a \sin \theta \quad \text{or} \quad r = b + a \cos \theta$$

is called a **limaçon**. The shape of the graph depends on the relative values of a and b. If $a = b$, the limaçon is called a cardioid from its heartlike shape, as illustrated in Figure 12. If the absolute value of b is greater than the absolute value of a, the graph is a curve surrounding the origin (see Fig. 9). An interesting feature is introduced in the graph when absolute value of a is greater than the absolute value of b. The graph then has an inner loop, which we illustrate in an example.

Example 1. Construct the graph of the limaçon

$$r = 2 + 4 \cos \theta$$

Solution. The equation is unchanged when θ is replaced by $-\theta$, since $\cos(-\theta) = \cos \theta$. Hence the graph is symmetric with respect to the polar axis. Setting $r = 0$ gives

$$2 + 4 \cos \theta = 0$$

$$\cos \theta = -\tfrac{1}{2}$$

$$\theta = 120°, 240°$$

Hence the lines $\theta = 120°$ and $\theta = 240°$ are tangent to the graph at the origin. To obtain further aid in constructing the graph, we prepare the following diagram.

θ	$\cos \theta$	r
$0° \rightarrow 90°$	$1 \rightarrow 0$	$6 \rightarrow 2$
$90° \rightarrow 120°$	$0 \rightarrow -\tfrac{1}{2}$	$2 \rightarrow 0$
$120° \rightarrow 180°$	$-\tfrac{1}{2} \rightarrow -1$	$0 \rightarrow -2$

The graph is shown in Fig. 15. The lower half of the large loop and the upper half of the small loop were drawn by the use of symmetry.

The graphs of the polar equations

$$r^2 = a^2 \sin 2\theta \qquad \text{and} \qquad r^2 = a^2 \cos 2\theta$$

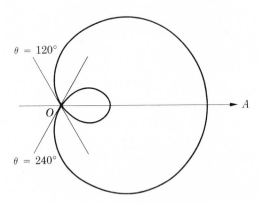

$\theta = 120°$

$\theta = 240°$

O

A

Fig. 15. $r = 2 + 4 \cos \theta$

are **lemniscates**. In each of these equations r varies from $-a$ to a, and the values of θ which make the right member negative are excluded. The excluded values in the first equation are $90° < \theta < 180°$ and $270° < \theta < 360°$. In the second equation the excluded values are $45° < \theta < 135°$ and $225° < \theta < 315°$.

Example 2. Draw the graph of the equation $r^2 = 9 \cos 2\theta$.

Solution. We observe that the equation is symmetric with respect to the pole, the polar axis, and the vertical line through the pole. As θ increases from $0°$ to $45°$, the positive values of r vary from 3 to 0 and the negative values from -3 to 0. Hence this interval for θ gives rise to the half loop in the first quadrant and the half loop in the third quadrant. Either of these half loops combined with the known symmetries is sufficient for completing the graph (Fig. 16). Like this graph, except for position, is the graph of $r^2 = 9 \sin 2\theta$ (Fig. 17).

Finally, the graphs of the equations $r = a\theta$, $r\theta = a$, and $r = e^{a\theta}$ are examples of **spirals**. The graphs of the first two equations are shown in Figs. 18 and 19. We consider the third equation in an example.

Example 3. Construct the graph of the equation $r = e^{0.2\theta}$.

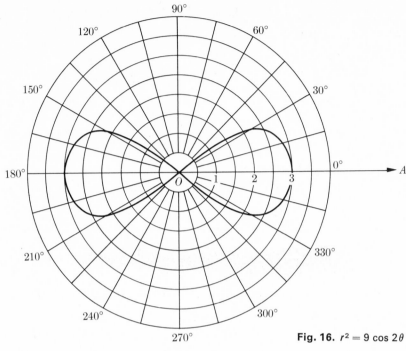

Fig. 16. $r^2 = 9 \cos 2\theta$

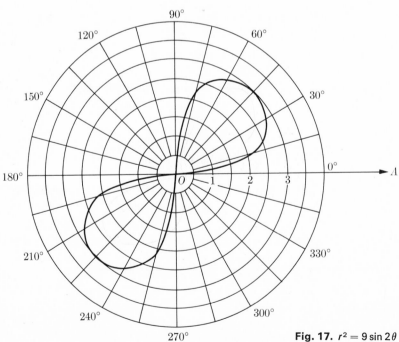

Fig. 17. $r^2 = 9 \sin 2\theta$

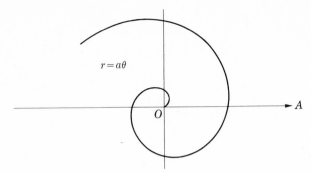

Fig. 18. Spiral of Archimedes

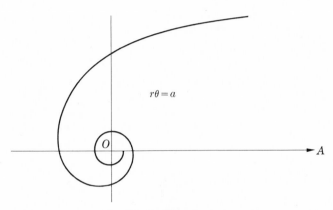

Fig. 19. Reciprocal spiral

Solution. Using Table III in the Appendix, we prepare a table of corresponding values of θ and r and obtain the curve in Fig. 20. The curve is called a **logarithmic spiral.**

θ	0	$\dfrac{\pi}{4}$	$\dfrac{\pi}{2}$	$\dfrac{3\pi}{4}$	π	$\dfrac{5\pi}{4}$	$\dfrac{3\pi}{2}$
0.2θ	0	0.157	0.314	0.471	0.628	0.786	0.942
r	1	1.17	1.37	1.60	1.88	2.20	2.57

θ	$\dfrac{7\pi}{4}$	2π	$\dfrac{9\pi}{4}$	$\dfrac{5\pi}{2}$	$\dfrac{11\pi}{4}$	3π
0.2θ	1.10	1.26	1.41	1.57	1.73	1.89
r	3.00	3.53	4.10	4.81	5.64	6.63

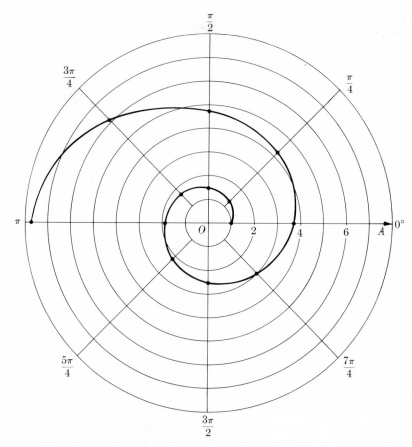

Fig. 20. $r = e^{0.2\theta}$

EXERCISES

1. Observe that (r, θ) and $(-r, 180° - \theta)$ are symmetric with respect to the polar axis, and that (r, θ) and $(-r, -\theta)$ are symmetric with respect to the line $\theta = 90°$. On the basis of this information, state two tests for the symmetry of the graph of an equation. Apply the tests to the equation $r = \sin 2\theta$.

 Sketch the graph of each of the following equations. First examine the equation to find properties which are helpful in tracing the graph. Where the literal constant a occurs, assign to it a convenient positive value. In the spirals 25 through 29 use radian measure for θ.

2. $r = 4(1 - \cos \theta)$ 3. $r = 6(1 - \sin \theta)$

4. $r = a(1 + \cos \theta)$ 5. $r = 5 - 2 \sin \theta$

6. $r = 10 - 5 \cos \theta$ 7. $r = 8 + 4 \cos \theta$

8. $r = 8 \cos 2\theta$

9. $r = a \sin 2\theta$

10. $r = 6 \sin 3\theta$

11. $r = 4 \cos 3\theta$

12. $r = 2 \sin 5\theta$

13. $r = 2 \cos 5\theta$

14. $r = 4 - 8 \cos \theta$

15. $r = 6 - 3 \sin \theta$

16. $r = 4 + 8 \cos \theta$

17. $r = 6 + 3 \sin \theta$

18. $r = a \cos 4\theta$

19. $r = a \sin 4\theta$

20. $r^2 = 16 \cos 2\theta$

21. $r^2 = 16 \sin 2\theta$

22. $r^2 = -a^2 \cos 2\theta$

23. $r^2 = -a^2 \sin 2\theta$

24. $r^2 = a^2 \cos\theta$

25. $r = 2\theta$

26. $r\theta = 4$

27. $r = e^\theta$

28. $r^2\theta = a$ (lituus)

29. $r^2 = a^2\theta$ (parabolic spiral)

30. $r = \sin \tfrac{1}{2}\theta$

31. $r = \cos \tfrac{1}{2}\theta$

32. $r = 3 \sec \theta + 4$

33. $r = 6 \sec \theta - 6$

6. POLAR EQUATIONS OF LINES AND CIRCLES

The equations of lines and circles can be obtained in polar equations by trans-
forming the rectangular coordinate equations of these curves. The polar
equations can also be derived directly. We shall derive the polar equation of a
line, and also of a circle, in general positions and in certain special positions.

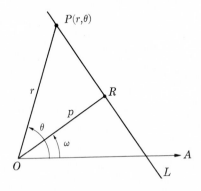

Figure 21

In Fig. 21 the segment OR is drawn perpendicular to the line L. We denote
the length of this segment by p and the angle which it makes with the polar

axis by ω. The coordinates of a variable point on the line are (r, θ). From the right triangle ORP, we have

$$\frac{p}{r} = \cos(\theta - \omega)$$

or

$$\boxed{r \cos(\theta - \omega) = p} \tag{5}$$

This equation holds for all points of the line. If P is chosen below OA, then the angle ROP is equal to $(\omega + 2\pi - \theta)$. Although this angle is not equal to $(\theta - \omega)$, we do have

$$\cos(\omega + 2\pi - \theta) = \cos(\omega - \theta) = \cos(\theta - \omega)$$

In a similar way, the equation could be derived for the line L in any other position and not passing through the origin.

Equation (5) is called the **polar normal form** of the equation of a straight line. For lines perpendicular to the polar axis $\omega = 0°$ or $180°$, and for lines parallel to the polar axis $\omega = 90°$ or $270°$. Substituting these values for ω, we have the special forms

$$\boxed{r \cos \theta = \pm p} \tag{6}$$

and

$$\boxed{r \sin \theta = \pm p} \tag{7}$$

The θ-coordinate is constant for points on a line passing through the origin. Hence the equation of a line through the origin with inclination α is

$$\boxed{\theta = \alpha} \tag{8}$$

Although the equation of a line through the origin can be written immediately in this form, it is worth noting that Eq. (8) is a special case of Eq. (5). By setting $p = 0$ in Eq. (5), we have $r \cos(\theta - \omega) = 0, \cos(\theta - \omega) = 0, \theta - \omega = \frac{1}{2}\pi$, and $\theta = \frac{1}{2}\pi + \omega = \alpha$.

It is worth noting that Eqs. (6) and (7) can be obtained directly from figures. The line L in Fig. 22 is perpendicular to the polar axis and p units to the right of the pole. For any point $P(r, \theta)$ on the line, $\cos \theta = p/r$ and therefore $r \cos \theta = p$. A line perpendicular to the polar axis and p units to the left of the pole is represented by the equation $r \cos \theta = -p$. Thus, the graph of $r \cos \theta = -4$ is the line

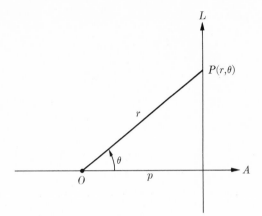

Figure 22

perpendicular to the polar axis and 4 units to the left of the pole. Similarly, the graph of $r\sin\theta = 4$ is the line parallel to the polar axis and 4 units above the pole, and the graph of $r\sin\theta = -4$ is the line parallel to the polar axis and 4 units below the pole.

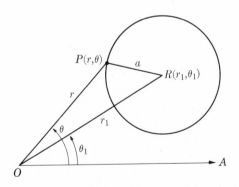

Figure 23

We next write the equation of a circle of radius a and center at $R(r_1,\theta_1)$. Observing Fig. 23 and applying the law of cosines to the triangle ORP, we get the equation of the circle in the form

$$r^2 + r_1^2 - 2rr_1\cos(\theta - \theta_1) = a^2 \qquad (9)$$

If the center is at $(a, 0°)$, then $r_1 = a$ and $\theta_1 = 0°$. The equation then reduces to

$$r = 2a \cos \theta \qquad (10)$$

If the center is at $(a, 90°)$, the equation becomes

$$r = 2a \sin \theta \qquad (11)$$

Example 1. If $\omega = 30°$ and $p = 2$ in Eq. (5), find the coordinates of the points at which the line intersects the polar axis and the vertical line $\theta = 90°$.

Solution. We replace θ by $0°$ in the equation $r \cos(\theta - 30°) = 2$ to find the intercept on the polar axis, and replace θ by $90°$ to find the intercept on the vertical line. This gives, respectively, $4\sqrt{3}/3$ and 7. Hence the required coordinates are $\left(\dfrac{4\sqrt{3}}{3}, 0°\right)$ and $(4, 90°)$. The line, of course, passes through these points.

Example 2. Find the equation of the circle with center at $(5, 60°)$ and radius 3.
Solution. Substituting in Eq. (9), we have

$$r^2 + 25 - 10r \cos (\theta - 60°) = 9$$

$$r^2 - 10r \cos (\theta - 60°) + 16 = 0$$

EXERCISES

1. From a figure find the equation of the line perpendicular to the polar axis and (a) 3 units to the right of the pole, (b) 3 units to the left of the pole. Compare your results with formula (6).

2. From a figure find the equation of the line parallel to the polar axis and (a) 3 units below the axis, (b) 3 units above the axis. Check your results with formula (7).

3. Show that the equation $Ax + By = C$, when expressed in polar form, becomes

$$r = \frac{C}{A \cos \theta + B \sin \theta}$$

   Show also that formula (5) can be reduced to this form.

Assign convenient values to θ and find the coordinates of two points on the line represented by each equation in Exercises 4 through 9. Plot the points and draw the line.

4. $r = \dfrac{1}{\cos\theta + 2\sin\theta}$

5. $r = \dfrac{4}{2\cos\theta - \sin\theta}$

6. $r = \dfrac{-10}{2\cos\theta + 5\sin\theta}$

7. $r = \dfrac{-12}{3\cos\theta - 2\sin\theta}$

8. $r = \dfrac{3}{\sin\theta - 6\cos\theta}$

9. $r = \dfrac{-4}{2\sin\theta + 3\cos\theta}$

Write the polar equation of each line described in Exercises 10 through 15.

10. The horizontal line through the point $(2, 90°)$.

11. The horizontal line through the point $(-2, 90°)$.

12. The vertical line through the point $(-4, 0°)$.

13. The vertical line through the point $(3, -\pi)$.

14. The line tangent to the circle $r = 2$ at the point $(2, 60°)$.

15. The line tangent to the circle $r = 4$ at the point $(4, 225°)$.

Give the polar coordinates of the center and the radius of the circle defined by each of Equations 16 through 21.

16. $r = 8\cos\theta$

17. $r = 6\sin\theta$

18. $r = -10\sin\theta$

19. $r = -4\cos\theta$

20. $r = 12\cos\theta$

21. $r = -7\sin\theta$

Find the polar equation of the circle in each of Exercises 22 through 30.

22. Center at $(3, 0°)$ and radius 3.

23. Center at $(4, 0°)$ and radius 4.

24. Center at $(6, 90°)$ and radius 6.

25. Center at $(-5, 0°)$ and radius 5.

26. Center at $(-8, 90°)$ and radius 8.

27. Center at $(4, 0°)$ and radius 2.

28. Center at $(4, 90°)$ and radius 2.

29. Center at $(5, 45°)$ and radius 5.

30. Center at $(6, 60°)$ and radius 6.

31. Center at $(3, 120°)$ and radius 2.

32. Center at $(7, 240°)$ and radius 6.

33. Derive formulas (10) and (11) directly from figures. You may use the condition that any angle inscribed in a semicircle is a right angle.

7. POLAR EQUATIONS OF CONICS

We use the focus-directrix property of conics (Chapter 3, Section 6) to derive polar equations of conics. The equations can be obtained in simple forms if a focus is at the origin and the directrix is parallel or perpendicular to the polar axis. In Fig. 24 the directrix D is perpendicular to the polar axis and to the

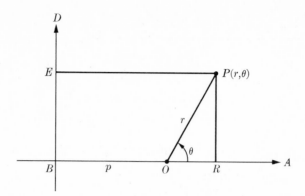

Figure 24

left of the origin. We indicate the eccentricity by e and the length of BO by p. Then for any point $P(r, \theta)$ of the conic, we have, by definition,

$$\frac{|OP|}{|EP|} = e$$

But the numerator $|OP| = r$ and the denominator

$$|EP| = |BR| = |BO| + |OP| \cos \theta = p + r \cos \theta$$

Hence $r/(p + r \cos \theta) = e$, and solving for r, we get

$$r = \frac{ep}{1 - e \cos \theta} \tag{12}$$

If a focus is at the pole and the directrix is p units to the right of the pole, the equation is

$$r = \frac{ep}{1 + e \cos \theta} \tag{13}$$

If a focus is at the pole and the directrix D is parallel to the polar axis and p units above the axis, then we have (Fig. 25)

$$\frac{|OP|}{|EP|} = e$$

We observe that $|OP| = r$ and $|EP| = p - |PR| = p - r \sin \theta$.

Hence $r/(p - r \sin \theta) = e$ and, solving for r, we get

$$r = \frac{ep}{1 + e \sin \theta} \qquad (14)$$

If a focus is at the pole and the directrix is p units below the polar axis, the equation is

$$r = \frac{ep}{1 - e \sin \theta} \qquad (15)$$

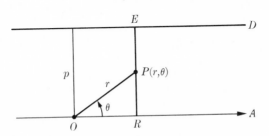

Figure 25

An equation in any of the forms (12) through (15) represents a parabola if $e = 1$, an ellipse if e is between 0 and 1, and a hyperbola if e is greater than 1. In any case the graph can be sketched immediately. Having observed the type of conic from the value of e, the next step is to find the points where the curve cuts the polar axis, the extension of the axis through O, and the line through the pole perpendicular to the polar axis. These are called the **intercept points,** and may be obtained by using the values 0°, 90°, 180°, and 270° for θ. Only three of these values can be used for a parabola, since one of them would make the denominator zero. The intercept points are sufficient for a rough graph. For increased accuracy a few additional points should be plotted.

Example 1. Sketch the graph of the equation $r = 8/(3 + 3 \cos \theta)$.

Solution. We divide the numerator and denominator of the fraction by 3 and have

$$r = \frac{8/3}{1 + \cos \theta}$$

This equation is of the form (13) with $e = 1$. Hence the graph is a parabola with axis along the extension of the polar axis. By substituting $0°$, $90°$, and $270°$ in succession for θ, we find the intercept points to be

$$(\tfrac{4}{3}, 0°) \qquad (\tfrac{8}{3}, 90°) \qquad (\tfrac{8}{3}, 270°)$$

The first of these points is the vertex of the parabola and the second and third are the ends of the latus rectum. Substituting $\theta = 60°$ and $\theta = 120°$ in the given equation and noting that the graph is symmetric with respect to its axis, we find the additional points

$$(\tfrac{16}{9}, 60°) \qquad (\tfrac{16}{3}, 120°) \qquad (\tfrac{16}{3}, 240°) \qquad (\tfrac{16}{9}, 300°)$$

These points, along with the intercept points, permit us to draw the graph (Fig. 26).

Example 2. Sketch the graph of $r = 15/(3 - 2\cos\theta)$.

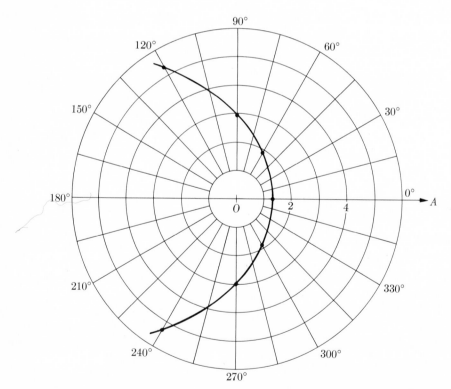

Fig. 26. $r = \dfrac{8}{3 + 3\cos\theta}$

Solution. The equation takes the form of Eq. (12) when the numerator and denominator of the right member are divided by 3. This produces

$$r = \frac{5}{1 - (\frac{2}{3}) \cos \theta}$$

In this form we observe that $e = \frac{2}{3}$, and hence the graph is an ellipse. Substituting $0°$, $90°$, $180°$, and $270°$ in succession for θ in the original equation, we find the intercept points to be

$$(15, 0°) \qquad (5, 90°) \qquad (3, 180°) \qquad (5, 270°)$$

These points are plotted in Fig. 27. The points $(15, 0°)$ and $(3, 180°)$ are vertices and the other intercept points are the ends of a latus rectum. The center, midway betweeen the vertices, is at $(6, 0°)$. Then, denoting the distance from

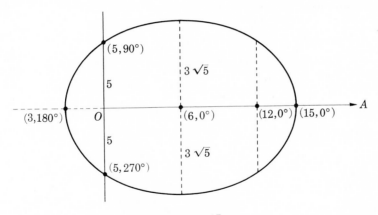

Fig. 27. $r = \dfrac{15}{3 - 2 \cos \theta}$

the center to a vertex by a, the distance from the center to a focus by c, and the distance from the center to an end of the minor axis by b, we have

$$a = 9 \qquad c = 6 \qquad b = \sqrt{a^2 - c^2} = 3\sqrt{5}$$

Example 3. Sketch the graph of the equation $r = 4/(2 + 3 \sin \theta)$.

Solution. Dividing the numerator and denominator of the fraction by 2, we have

$$r = \frac{2}{1 + 1.5 \sin \theta}$$

This equation is in the form (14) with $e = 1.5$. Hence the graph is a hyperbola. Although the graph could be roughly sketched from the intercepts, we shall plot a few additional points by use of a table of values.

θ	0°	30°	90°	150°	180°	205°	270°	335°
r	2	$\frac{8}{7}$	$\frac{4}{5}$	$\frac{8}{7}$	2	5.5	-4	5.5

The point $(0.8, 90°)$ is the vertex of the lower branch of the hyperbola and the point $(-4, 270°)$ is the vertex of the upper branch. The center of the hyperbola, midway between the vertices, is at the point $(2.4, 90°)$. The denominator of the given fraction is equal to zero when $\sin\theta = -\frac{2}{3}$. If we denote this angle by $180° + \theta_0$, then for all values of θ such that $180° + \theta_0 < \theta < 360° - \theta_0$, the values of r are negative and yield the points on the upper branch. We can, of course, draw the upper branch by noticing that the hyperbola is symmetric with respect to its center $(2.4, 90°)$. The graph is shown in Fig. 28.

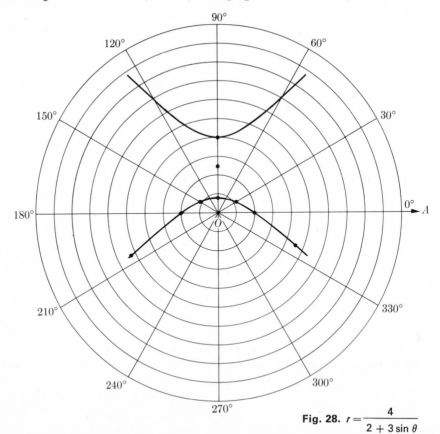

Fig. 28. $r = \dfrac{4}{2 + 3\sin\theta}$

EXERCISES

Express each of the following equations in one of the forms (12) through (15) and sketch the graph of the conic.

1. $r = \dfrac{4}{1 - \cos \theta}$

2. $r = \dfrac{6}{1 + \sin \theta}$

3. $r = \dfrac{9}{2 + 2 \cos \theta}$

4. $r = \dfrac{10}{3 - 3 \sin \theta}$

5. $r = \dfrac{12}{2 - \cos \theta}$

6. $r = \dfrac{12}{2 + \sin \theta}$

7. $r = \dfrac{16}{4 + 3 \cos \theta}$

8. $r = \dfrac{15}{5 - 4 \sin \theta}$

9. $r = \dfrac{2}{1 - 2 \cos \theta}$

10. $r = \dfrac{4}{2 + 3 \cos \theta}$

11. $r = \dfrac{8}{3 + 5 \sin \theta}$

12. $r = \dfrac{6}{3 - 4 \sin \theta}$

8. INTERSECTIONS OF POLAR COORDINATE GRAPHS

A simultaneous real solution of two equations in rectangular coordinates represents a point of intersection of their graphs. Conversely, the coordinates of a point of intersection yield a simultaneous solution. In polar coordinates, however, this converse statement does not always hold. This difference in the two systems is a consequence of the fact that a point has more than one pair of polar coordinates. As an illustration, consider the equations $r = -2$, $r = 1 + \sin \theta$ and the two pairs of coordinates $(2, 90°)$, $(-2, 270°)$. The equation $r = -2$ is satisfied by the second pair of coordinates but not by the first. The equation $r = 1 + \sin \theta$ is satisfied by the first pair of coordinates but not by the second. The two pairs of coordinates, however, determine the same point. Although the two curves pass through this point, no pair of coordinates of the point satisfies both equations. The usual process of solving two equations simultaneously does not yield an intersection point of this kind. The graphs of the equations, of course, show all intersections.

Example 1. Solve simultaneously and sketch the graphs of

$$r = 6 \sin \theta \quad \text{and} \quad r = 6 \cos \theta$$

Solution. Equating the right members of the equations, we have

$$6 \sin \theta = 6 \cos \theta$$

$$\tan \theta = 1$$

$$\theta = 45°, 225°$$

$$r = 3\sqrt{2}, -3\sqrt{2}$$

The coordinates $(3\sqrt{2}, 45°)$ and $(-3\sqrt{2}, 225°)$ define the same point. The graphs (Fig. 29) show this point, and show also that both curves pass through

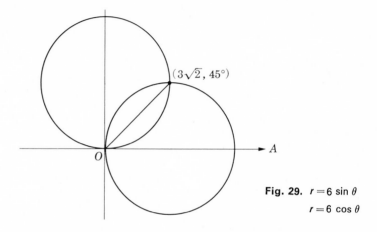

Fig. 29. $r = 6 \sin \theta$
$r = 6 \cos \theta$

the origin. The coordinates $(0, 0°)$ satisfy the first equation and $(0, 90°)$ satisfy the second equation. But the origin has no pair of coordinates which satisfies both equations.

Example 2. Solve simultaneously and draw the graphs of

$$r = 4 \sin \theta \qquad \text{and} \qquad r = 4 \cos 2\theta$$

Solution. Eliminating r and using the trigonometric identity $\cos 2\theta = 1 - 2 \sin^2 \theta$, we obtain

$$4 \sin \theta = 4(1 - 2 \sin^2 \theta)$$

$$2 \sin^2 \theta + \sin \theta - 1 = 0$$

$$(2 \sin \theta - 1)(\sin \theta + 1) = 0$$

$$\sin \theta = \tfrac{1}{2}, -1$$

$$\theta = 30°, 150°, 270°$$

$$r = 2, \qquad 2, \qquad -4$$

The solutions are $(2, 30°)$, $(2, 150°)$, and $(-4, 270°)$. Figure 30 shows that the curves also cross at the origin, but the origin has no pair of coordinates which satisfies both equations.

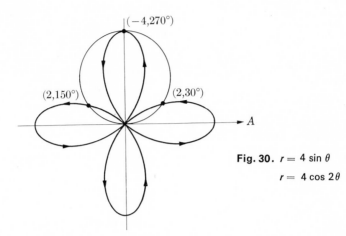

Fig. 30. $r = 4 \sin \theta$
$r = 4 \cos 2\theta$

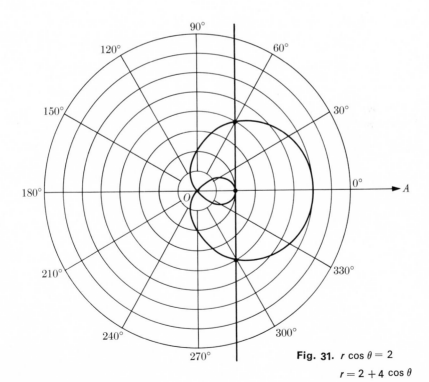

Fig. 31. $r \cos \theta = 2$
$r = 2 + 4 \cos \theta$

Example 3. Solve simultaneously and sketch the graphs of the equations

$$r \cos \theta = 2 \quad \text{and} \quad r = 2 + 4 \cos \theta$$

Solution. Eliminating r between the equations and simplifying the result, we obtain

$$2 \cos^2 \theta + \cos \theta - 1 = 0 \quad \text{or} \quad (2 \cos \theta - 1)(\cos \theta + 1) = 0$$

We see from the last equation that $\theta = 60°$, $\theta = 300°$, and $\theta = 180°$ are solutions. The corresponding values of r are, respectively, $r = 4$, $r = 4$, $r = -2$. As may be verified, the coordinates of the points $(4, 60°)$, $(4, 300°)$, and $(-2, 180°)$ satisfy both of the given equations. The graphs are shown in Fig. 31.

EXERCISES

In each of the following Exercises solve the equations simultaneously and sketch their graphs. Extraneous solutions are sometimes introduced in the solving process. For this reason all results should be checked.

1. $r = 2 \cos \theta$
   $r = 1$

2. $r = 4 \sin \theta$
   $r = 4 \cos \theta$

3. $r = 6 \cos \theta$
   $r \cos \theta = 3$

4. $r = a(1 + \sin \theta)$
   $r = 2a \sin \theta$

5. $r = a(1 + \cos \theta)$
   $r = a(1 - \cos \theta)$

6. $r \cos \theta = 1$
   $r = 4 \cos \theta$

7. $r^2 = 4 \cos \theta$
   $r = 2$

8. $r = 1 + \sin \theta$
   $r = 1 + \cos \theta$

9. $r = \dfrac{2}{1 + \cos \theta}$
   $3r \cos \theta = 2$

10. $r = \dfrac{3}{4 - 3 \cos \theta}$
    $r = 3 \cos \theta$

11. $r = 2 \sin \theta + 1$
    $r = \cos \theta$

12. $r^2 = a^2 \sin 2\theta$
    $r = a\sqrt{2} \cos \theta$

13. $r = 4 \cos \theta$
    $r = 4 \sin 2\theta$

14. $r = \sin^2 \theta$
    $r = \cos^2 \theta$

15. $r = 2 \sin \tfrac{1}{2}\theta$
    $r = 1$

16. $r = 1 - \sin \theta$
    $r = \cos 2\theta$

17. $r = 4 + \cos \theta$
    $r \cos \theta = -3$

18. $r = 4 - \sin \theta$
    $r \sin \theta = 3$

19. $r = 2 \cos \theta + 1$
    $r \cos \theta = 1$

20. $r = \dfrac{2}{\sin \theta + \cos \theta}$
    $r = \dfrac{2}{1 - \cos \theta}$

Chapter 6

Parametric Equations

1. PARAMETRIC REPRESENTATION OF CURVES

We have obtained graphs of equations in the two variables x and y. Another way of defining a graph is to express x and y separately in terms of a third variable. Equations of this kind are of much importance as the mathematical treatment of many problems is facilitated by their use. Before illustrating the two-equation situation, we state the following definition.

DEFINITION 1 *The equation of a graph in two dimensions is said to be in* ***parametric form*** *if each coordinate of a general point $P(x, y)$ is expressed in terms of a third variable. The third variable, usually denoted by a letter, is called a* ***parameter***.

The equations

$$x = t - 1 \qquad \text{and} \qquad y = 2t + 3$$

for example, are parametric equations, and t is the parameter. The equations define a graph. If t is assigned a value, corresponding values are determined for x and y. The pair of values for x and y constitute the coordinates of a point of the graph. The complete graph consists of the set of all points determined in this way as t varies through all its chosen values. We can eliminate t between the equations and obtain an equation involving x and y. Thus, solving either equation for t and substituting in the other, we get

$$2x - y + 5 = 0$$

The graph of this equation, which is also the graph of the parametric equations, is a straight line.

Often the parameter can be eliminated, as illustrated here, to obtain an equation in x and y. Sometimes, however, the process is not easy or possible because the parameter is included in a complicated way. The equations

$$x = t^5 + \log t \qquad \text{and} \qquad y = t^3 + \tan t$$

illustrate this statement.

It is sometimes helpful in solving a problem to change an equation in x and y to parametric form. We illustrate this process with the equation

$$x^2 + 2x + y = 4,$$

which defines a parabola. If we substitute $2t$ for x and solve the resulting equation for y, we get $y = 4 - 4t - 4t^2$. Hence the parametric equations

$$x = 2t \qquad \text{and} \qquad y = 4 - 4t - 4t^2$$

also represent the parabola. It is evident that other representations could be obtained by equating x to other expressions in t. Again, this procedure is inconvenient or perhaps impossible in equations which contain both variables in a complicated way.

The parameter as used in this chapter plays a different role from the parameter in Chapter 2, Sections 4 and 7. Here a curve is determined by letting the parameter vary. In the earlier use the parameter gave rise to a set of curves (lines and circles).

2. PARAMETRIC EQUATIONS OF THE CIRCLE, ELLIPSE, AND HYPERBOLA

We have already illustrated parametric representations of a straight line and parabola. We turn now to a consideration of the circle and the remaining conics.

To find a parametric representation of the circle of radius a and center at the origin, we select for the parameter the angle θ, as indicated in Fig. 1. Recalling the definitions of the sine and cosine of an angle, we have

$$\frac{x}{a} = \cos \theta \qquad \text{and} \qquad \frac{y}{a} = \sin \theta$$

or, equivalently, the pair of equations

$$x = a \cos \theta \qquad y = a \sin \theta$$

If we let θ increase from $0°$ to $360°$, the point $P(x, y)$ defined by these equations starts at $(a, 0)$ and moves counterclockwise around the circle.

Next, let the center of a circle be at (h, k) with radius a (Fig. 2). For the directed distances $\overrightarrow{CQ}$ and $\overrightarrow{QP}$, we obtain

$$\overrightarrow{CQ} = x - h = a \cos \theta \qquad \text{and} \qquad \overrightarrow{QP} = y - k = a \sin \theta$$

or the pair of equations

$$x = h + a \cos \theta$$

$$y = k + a \sin \theta$$

If a and b are any unequal positive numbers, the equations $x = a \cos \theta$ and $y = b \sin \theta$ represent an ellipse. This statement can be verified by eliminating the parameter θ. Writing the equations as $x/a = \cos \theta$ and $y/b = \sin \theta$, squaring the members of each equation and adding, we get

$$\left(\frac{x}{a}\right)^2 + \left(\frac{y}{b}\right)^2 = \cos^2 \theta + \sin^2 \theta$$

or

$$\frac{x^2}{a^2} + \frac{y^2}{b^2} = 1$$

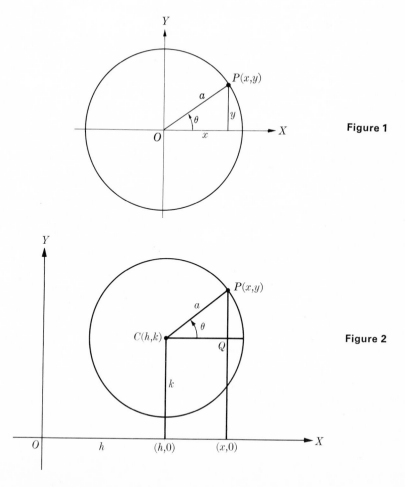

Figure 1

Figure 2

From this result we see that the parametric equations represent an ellipse with a and b as semiaxes. The geometric significance of θ can be determined by referring to Fig. 3. The radius of the smaller circle is b and the radius of the larger circle is a. The terminal side of θ cuts the circles at B and A. The horizontal line

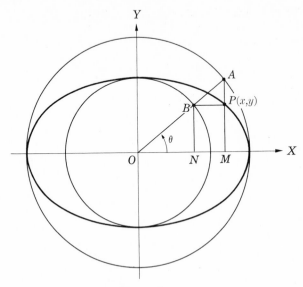

Figure 3

through B and the vertical line through A intersect at $P(x, y)$. For this point, we have

$$x = \overrightarrow{OM} = |OA| \cos \theta = a \cos \theta$$

$$y = \overrightarrow{MP} = \overrightarrow{NB} = |OB| \sin \theta = b \sin \theta$$

Hence $P(x, y)$ is a point of the ellipse. As θ varies, P moves along the ellipse. If θ starts at $0°$ and increases to $360°$, the point P starts at $(a, 0)$ and traverses the ellipse in a counterclockwise direction.

Let a and b be positive numbers, and consider the parametric equations

$$x = a \sec \theta \qquad \text{and} \qquad y = b \tan \theta$$

Since $\sec^2 \theta - \tan^2 \theta = 1$, we may write

$$\left(\frac{x}{a} \right)^2 - \left(\frac{y}{b} \right)^2 = \sec^2 \theta - \tan^2 \theta$$

or

$$\frac{x^2}{a^2} - \frac{y^2}{b^2} = 1$$

We note that $\sec\theta$ and $\tan\theta$ exist for all angles except those for which θ is an odd multiple of $90°$. Suppose we let θ take all values such that $0° \leq \theta < 90°$. At $0°$, $\sec\theta = 1$ and $\tan\theta = 0$. As θ increases in this specified interval, $\sec\theta$ starts at 1 and assumes all positive values greater than 1, and $\tan\theta$ starts at 0 and assumes all positive values. Hence, x starts at a and assumes all positive values greater than a, and y starts at 0 and assumes all positive values. We see, then, that this chosen interval for θ provides for the portion of the hyperbola in the first quadrant. Similarly, the values of θ such that $90° < \theta \leq 180°$ represents the portion of the hyperbola in the third quadrant. The student may continue the discussion for the remaining quadrants.

3. GRAPHS OF PARAMETRIC EQUATIONS

We proceed next to the problem of constructing the graph defined by two parametric equations. The method is straightforward. We first assign to the parameter a set of values and compute the corresponding values of x and y. The plotted points (x, y) furnish a guide for drawing the graph. Usually only a few plotted points are necessary. This is especially true when certain properties of the graph, such as the extent, the intercepts, and the symmetry are apparent from the equations.

If a pair of parametric equations is to represent a rectangular equation, it is well to determine if the equations and the domain of the parameter are suitably chosen. That is, if the coordinates $P(x, y)$ satisfy the rectangular equation, there must be a value of the parameter so that the parametric equations yield the same coordinates. Conversely, any pair of coordinates obtained from the parametric equations must satisfy the rectangular equation. In some cases, the graph of the parametric equations and the graph of the corresponding rectangular equation do not coincide throughout. Example 2 below illustrates such a case.

Example 1. Sketch the graph of the parametric equations

$$x = 2 + t \qquad \text{and} \qquad y = 3 - t^2$$

Solution. By inspecting the equations, we see that x may have any real value and that y may have any real value not exceeding 3. We are able to sketch the curve (Fig. 4) from the accompanying table of values.

t	-3	-2	-1	0	1	2	3
x	-1	0	1	2	3	4	5
y	-6	-1	2	3	2	-1	-6

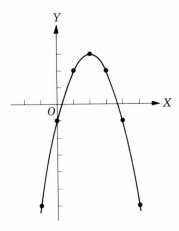

Figure 4

Example 2. Construct the graph of the equations

$$x = \cos^2 \theta \quad \text{and} \quad y = 2 \sin \theta$$

Solution. Let us eliminate the parameter θ from these equations. The second equation yields $y^2/4 = \sin^2 \theta$. Then adding the corresponding members of this equation and the first given equation, we obtain

$$\frac{y^2}{4} + x = 1$$

and consequently,

$$y^2 = -4(x - 1)$$

The graph of this equation is the parabola drawn in Fig. 5. We note, however, that the graph of the parametric equations does not include the part of the parabola to the left of the y axis. This results from the fact that the values of x are nonnegative.

It may happen that an equation in its original form is difficult to plot, and yet can be represented by more manageable equations in parametric form. We illustrate this situation in an example.

Example 3. Find a parametric representation of the equation

$$y^{2/3} + x^{2/3} = a^{2/3}$$

which is suitable for constructing the graph.

Solution. Solving the equation for $y^{2/3}$, we get

$$y^{2/3} = a^{2/3} - x^{2/3}$$

$$= a^{2/3}\left[1 - \left(\frac{x}{a}\right)^{2/3}\right]$$

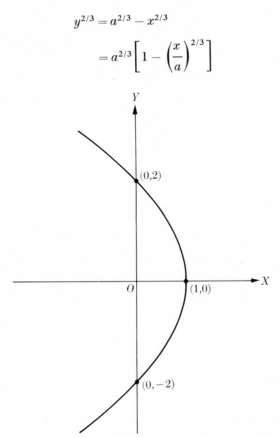

Figure 5

We observe that the bracketed expression may be simplified by setting $(x/a)^{2/3} = \sin^2\theta$ or $x = a\sin^3\theta$. When x has this value, we find that $y = a\cos^3\theta$. Hence the given equation is represented in parametric form by the equations

$$x = a\sin^3\theta$$

$$y = a\cos^3\theta$$

We can visualize the graph by letting θ increase, in 90° steps, from 0° to 360°. Thus, in the first step x increases from 0 to a and y decreases from a to 0. The graph (Fig. 6) is called a **hypocycloid of four cusps**. It can be shown that the path traced by a given point on a circle of radius $\frac{1}{4}a$ as it rolls inside and along a circle of radius a is a hypocycloid.

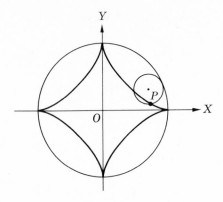

Figure 6

EXERCISES

In each of Exercises 1 through 4 sketch the graph by use of a table of values. Then compare the graph with that of the rectangular equation obtained by eliminating the parameter.

1. $x = 2t; y = 3t$

2. $x = 1 + t; y = 4 - 3t$

3. $x = t + 3; y = t^2 - 2$

4. $x = 1 + 2t; y = 2 - t^2$

Find the rectangular form of each pair of parametric equations. Then sketch the curve, using the simpler of the two forms.

5. $x = 5 \sin \theta; y = 5 \cos \theta$

6. $x = 2 + \sin \theta; y = 2 + \cos \theta$

7. $x = 4 \cos \theta; y = 3 \sin \theta$

8. $x = 2 \sec^2 \theta; y = 3 \tan^2 \theta$

9. $x = \cos 2\theta; y = 2 \sin \theta$

10. $x = \sec \theta; y = \tan \theta$

11. $x = -4 + 3 \cos \theta; y = 3 + 4 \sin \theta$

12. $x = 3 + 2 \cos \theta; y = -2 + 2 \sin \theta$

13. $x = \dfrac{2}{1 + t^2}; y = \dfrac{2t}{1 + t^2}$

14. $x = \dfrac{6t}{1 + t^2}; y = \dfrac{6t^2}{1 + t^2}$

Eliminate the parameter from each pair of equations. Draw the graph of the resulting equation and tell what part of the graph is covered by the parametric equations.

15. $x = 2 \sin^2 \theta; y = 3 \cos^2 \theta$

16. $x = \tan^2 \theta; y = \sec^2 \theta$

17. $x = 1 + 3 \sin^2 \theta; y = 1 + 3 \cos^2 \theta$

18. $x = \sin^2 \theta; y = 2 \cos \theta$

19. $x = 2 \cos^2 \theta; y = 4 \sin \theta$

20. $x = 6t^{-1}; y = t$

Using the accompanying equation, express each rectangular equation in parametric form

21. $2x - xy - 1 = 0$; $y = t + 2$ 22. $x^3 + y^3 - 3xy = 0$; $y = tx$

23. $x^{1/2} + y^{1/2} = a^{1/2}$; $x = a \sin^4 \theta$ 24. $x^2(y + 3) = y^3$; $x = ty$

Let t take values in steps of $\pi/2$ from 0 to 4π and find the corresponding values of x and y. Plot the points (x, y) thus determined and draw a smooth curve through the points by going from point to point in the order of increasing t.

25. $x = t \sin t$; $y = \cos t$ 26. $x = t \cos t$; $y = \sin t$

4. PATH OF A PROJECTILE*

The equations of certain curves can be determined more readily by the use of a parameter than otherwise. In fact, this is one of the principal uses of parametric equations. In the remainder of this chapter parametric equations of curves are required. These curves have interesting properties and also have important practical and theoretical applications.

We consider first the path of a projectile in air. Suppose that a body is given an initial upward velocity of v_0 feet per second in a direction which makes an angle α with the horizontal. If the resistance of the air is small and can be neglected without great error, the object will move subject to the vertical force of gravity. This means that there is no horizontal force to change the speed in the horizontal direction. Observing Fig. 7 with the origin of coordinates at the point where the projectile is fired, we see that the velocity in the x-direction is $v_0 \cos \alpha$. Then the distance traveled horizontally at the end of t seconds is $(v_0 \cos \alpha)t$ feet. Now the projectile is started with a vertical component of velocity of $v_0 \sin \alpha$ feet per second. This velocity would cause the projectile to rise upward to a height of $(v_0 \sin \alpha)t$ feet in t seconds. But the effect of the pull of gravity lessens this distance. According to a formula of physics the

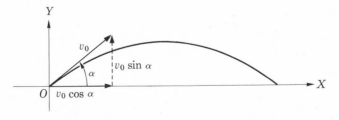

Figure 7

---

* If desired, the remainder of this chapter, or any part of it, may be omitted.

amount to be subtracted is $\frac{1}{2}gt^2$, where g is a constant and approximately equal to 32. Hence the parametric equations of the path are

$$x = (v_0 \cos \alpha)t \qquad y = (v_0 \sin \alpha)t - \tfrac{1}{2}gt^2 \tag{1}$$

If we solve the first equation for t and substitute the result in the second, we obtain the equation of the path in the rectangular form

$$y = (\tan \alpha)x - \frac{gx^2}{2v_0^2 \cos^2 \alpha} \tag{2}$$

This equation, which is of the second degree in x and the first degree in y, represents a parabola.

Example. A stone is thrown with a velocity of 160 ft/sec in a direction 45° above the horizontal. Find how far away the stone strikes the ground and its greatest height.

Solution. We substitute $v_0 = 160$, $\alpha = 45°$, and $g = 32$ in Eq. (1). This gives the parametric equations

$$x = 80\sqrt{2}\,t \qquad y = 80\sqrt{2}\,t - 16t^2$$

The stone reaches the ground when $y = 0$. We substitute this value for y in the second equation and find $t = 5\sqrt{2}$ sec as the time of flight. The value of x at this time is $x = 80\sqrt{2}(5\sqrt{2}) = 800$ ft. We know that the stone moves along a parabola which opens downward and that a parabola is symmetric with respect to its axis. Hence the greatest height is the value of y when t is half the flight time. Substituting $t = 5\sqrt{2}/2$ in the second equation, we find $y = 200$ ft. The stone strikes the ground 800 ft from the starting point and reaches a maximum height of 200 ft.

Alternatively, we can obtain the desired results by using the rectangular equation of the path. Thus substituting for v_0, α, and y in Eq. (2), we have

$$y = x - \frac{x^2}{800}$$

This equation, reduced to standard form, becomes

$$(x - 400)^2 = -800(y - 200)$$

The vertex, at $(400, 200)$, is the highest point. Letting $y = 0$, we find $x = 800$. Hence the stone strikes the ground at the point $(800, 0)$.

5. THE CYCLOID

The path traced by a given point on the circumference of a circle which rolls along a line is called a **cycloid.** To derive the equation of the cycloid, we select the line as the x axis and take the origin at a position where the tracing point is in contact with the x axis.

In Fig. 8 the radius of the rolling circle is a, and P is the tracing point. In the position drawn, the circle has rolled so that CP makes an angle θ (radians)

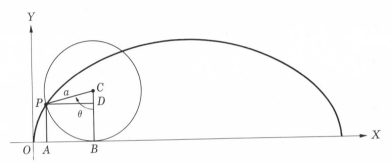

Figure 8

with the vertical. Since the circle rolls without slipping, the line segment OB and the arc PB are of equal length. Hence

$$\overrightarrow{OB} = \text{arc } \overrightarrow{PB} = a\theta$$

Observing the right triangle PDC, we may write

$$x = \overrightarrow{OA} = \overrightarrow{OB} - \overrightarrow{PD} = a\theta - a\sin\theta$$

$$y = \overrightarrow{AP} = \overrightarrow{BC} - \overrightarrow{DC} = a - a\cos\theta$$

The equations of the cycloid in parametric form are

$$x = a(\theta - \sin\theta) \qquad y = a(1 - \cos\theta)$$

The result of eliminating θ from these equations is the complicated equation

$$x = a\arccos\frac{a - y}{a} \pm \sqrt{2ay - y^2}$$

If a circle rolls beneath a line, a point of the circle would generate an inverted cycloid (Fig. 9). This curve has an interesting and important physical property. A body sliding without friction would move from A to B, two points on a downward part of the curve, in a shorter time than would be required along any other path connecting the two points. A proof of this property is too difficult to be included.

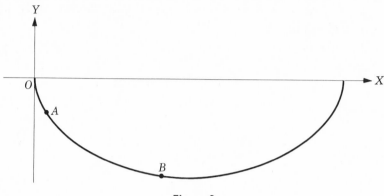

Figure 9

EXERCISES

In each of Exercises 1 through 4 write the parametric equations of the path of the object, using $g = 32$. Write also the rectangular equation of the path and give the requested information.

1. A ball is thrown with an initial velocity of 80 ft/sec and at an angle 45° above the horizontal. How high does the ball ascend and how far away, assuming the ground to be level, does it strike the ground?

2. A projectile is fired with an initial velocity of 160 ft/sec and at an angle 30° above the horizontal. Find the coordinates of its position at the end of (a) 1 sec, (b) 3 sec, (c) 5 sec. At what times is the projectile 64 ft above the starting point?

3. A projectile is fired horizontally ($\alpha = 0°$) from a building 64 ft high. If the initial velocity is v_0 ft/sec, find how far downward and how far horizontally the projectile travels in 2 sec.

4. A pitcher throws a baseball horizontally with an initial velocity of 108 ft/sec. If the point of release is 6 ft above the ground, at what height does the ball reach the home plate, 60.5 ft from the pitcher's box?

5. A circle of radius a rolls along a line. A point on a radius, b units from the center, describes a path. Paralleling the derivation in Section 5, show that the path is represented by the equations

$$x = a\theta - b\sin\theta \qquad y = a - b\cos\theta$$

The curve is called a **curtate cycloid** if $b < a$ and a **prolate cycloid** if $b > a$.

6. Sketch the curve of the equations in Exercise 5, taking $a = 4$ and $b = 3$. Sketch the curve if $a = 4$ and $b = 6$.

7. A circle of radius 4 rolls along a line and makes a revolution in 2 sec. A point, starting downward on a vertical radius, moves from the center to the circumference along the radius at a rate of 2 ft/sec. Find the equations of the path of the point.

8. The end of a thread kept in the plane of a circle describes a path called the **involute** of the circle, as it is unwound tautly from the circle. Use Fig. 10 to show that the parametric equations of the involute are

$$x = a(\cos\theta + \theta\sin\theta) \qquad y = a(\sin\theta - \theta\cos\theta)$$

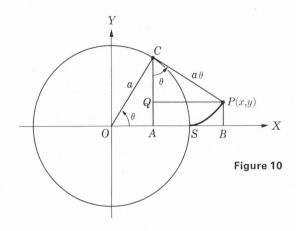

Figure 10

9. In Fig. 11 a circle of radius a is tangent to the two parallel lines OX and AC. The line OC cuts the circle at B, and $P\,(x, y)$ is the intersection of a

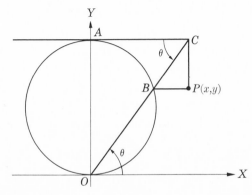

Figure 11

horizontal line through B and a vertical line through C. Show that the equations of the graph of P, as C moves along the upper tangent, are

$$x = 2a \cot \theta \qquad y = 2a \sin^2 \theta$$

This curve is called the **witch of Agnesi.** Show that its rectangular equation is

$$y = \frac{8a^3}{x^2 + 4a^2}$$

10. In Fig. 12, $OP = AB$. Show that the equations of the path traced by P, as A moves around the circle, are

$$x = 2a \sin^2 \theta \qquad y = 2a \sin^2 \theta \tan \theta$$

The curve is called the **cissoid of Diocles.** The rectangular equation is

$$y^2 = \frac{x^3}{2a - x}$$

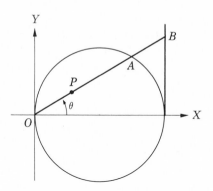

Figure 12

Chapter 7

Algebraic Curves

We considered first-degree equations in Chapter 2 and second-degree equations in Chapters 3 and 4. In this chapter we attack the problem of drawing the graphs of equations of a degree higher than two. All the equations define relations, and some of the equations define both functions and relations (Definitions 4 and 5, Chapter 1). The point-by-point method of constructing a graph is tedious except for simple equations. The task can often be lightened, however, by first discovering certain characteristics of the graph as revealed by the equation. We have already discussed the concepts of intercepts, symmetry, and extent of a curve. These topics occur in Chapter 1, Section 10, and Chapter 3, Sections 3 and 5, and should be restudied. The following section gives another helpful idea which we shall introduce before drawing the graphs of particular equations.

1. VERTICAL AND HORIZONTAL ASYMPTOTES

If the distance of a point from a straight line approaches zero as the point moves indefinitely far from the origin along a curve, then the line is called an **asymptote** of the curve. In drawing the graph of an equation it is well to determine the asymptotes, if any. They are often easily found and facilitate the graphing. The utility of the asymptotes in drawing a hyperbola has already been demonstrated (Chapter 3, Section 9). Here, however, we shall deal mainly with curves whose asymptotes, if any, are either horizontal or vertical.

Example 1. Examine the equation and draw the graph of

$$y = (x^2 - 1)(x - 2)^2$$

Solution. The three tests for symmetry (Chapter 3, Section 4) reveal that the graph does not have symmetry with respect to either axis or the origin. The x intercepts are -1, 1, and 2, and the y intercept is -4. These points greatly facilitate the graphing. As x increases beyond 2, the y values increase rapidly (Fig. 1). If x has a value between 1 and 2, the equation shows that y is positive

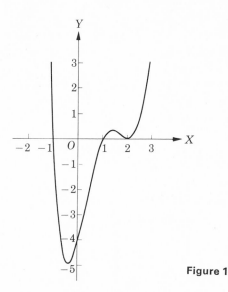

Figure 1

because both factors $x^2 - 1$ and $(x - 2)^2$ are then positive. If x has a value between -1 and 1, the factor $x^2 - 1$ is negative, and the curve is below the x axis. As x takes values to the left of $x = -1$, the y values become large. With the preceding information we can quickly sketch the curve. The plotting of a few points other than the intercepts makes possible a rather accurate graph.

Example 2. Examine the equation and draw the graph of

$$x^2 y - 4y = 8$$

Solution. The y intercept is -2. But if we set $y = 0$, there is obviously no value of x which will satisfy the equation. Hence there is no x intercept. The graph has symmetry with respect to the y axis but not with respect to the x axis. The part of the graph to the right of the y axis may first be determined and then the other drawn by the use of symmetry.

Solving the equation for y gives

$$y = \frac{8}{x^2 - 4} \tag{1}$$

We see that the right member of this equation is negative for $-2 < x < 2$, and consequently the graph is below the x axis in this interval. Furthermore, if x has a value slightly less than 2, the denominator is near zero. Then the fraction, which is equal to y, has a large absolute value. As x is assigned values less than but still closer to 2, the corresponding values of $|y|$ increase and can be made greater than any chosen number by taking x close enough to 2. This property of the given equation is indicated by the table of corresponding values of x and y. If, however, x is greater than 2, the value of y is positive. And y can be made to exceed any chosen positive number by letting x be greater than, yet close enough, to 2. Hence the line $x = 2$ is a vertical asymptote of the graph both below and above the x axis. To check for a horizontal asymptote, we think of assigning x larger and larger positive values in Eq. (1). It is evident that in this manner we can make y, though positive, as close to 0 as we wish. This means that $y = 0$ is an asymptote. The y intercept, the asymptotes, and the symmetric property aid greatly in constructing the graph of the given equation (Fig. 2).

x	0	1	1.5	1.9	1.99	1.999
y	-2	-2.7	-4.6	-20.5	-200	-2000

We may also determine the asymptotes in this example by solving the equation for x in terms of y. Thus, taking positive square roots, we find

$$x = 2\sqrt{\frac{y+2}{y}} \tag{2}$$

The radicand in this equation must not be negative. Hence the excluded values of y are $-2 < y < 0$. Observing Eq. (2), we see that a positive value of y close to 0 makes x large, and that we can make x as large as we please by taking y close enough to 0. Also, by taking larger and larger positive values for y, we can make the corresponding values of x as close to 2 as we wish. We conclude, then, that $y = 0$ and $x = 2$ are asymptotes of the graph.

The discussions in Examples 1 and 2 above should be helpful to the student in drawing the graphs of the equations in the Exercises at the end of Sections 1 and 2.

We next discuss an alternative procedure for determining horizontal asymptotes when y is equal to the quotient of two polynomials in x. Consider, for example, the equation

$$y = \frac{3x^3 - 2x^2 + x - 5}{2x^3 + 4x^2 - 8x - 1}$$

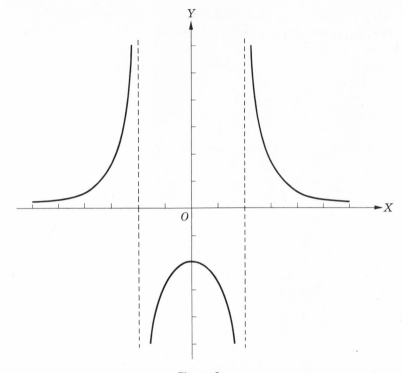

Figure 2

where the numerator and denominator are of the same degree. To determine the behavior of the right member as $|x|$ becomes large, we divide the numerator and denominator by x^3. This gives the equation

$$y = \frac{3 - (2/x) + (1/x^2) - (5/x^3)}{2 + (4/x) - (8/x^2) - (1/x^3)}$$

If now $|x|$ is assigned a large value, each term of both the numerator and denominator, after the first, is close to zero. Hence, the value of the fraction is close to $\frac{3}{2}$. Furthermore, the value can be made as near $\frac{3}{2}$ as desired by assigning $|x|$ a sufficiently large value. We conclude, then, that $y = \frac{3}{2}$ is an asymptote.

Suppose next that the degree of the numerator is less than the degree of the denominator, as in

$$y = \frac{3x^2 - 5x + 6}{2x^3 + 7x^2 + 5} = \frac{(3/x) - (5/x^2) + (6/x^3)}{2 + (7/x) + (5/x^3)}$$

Clearly, for a large $|x|$, the numerator is close to 0 and the denominator is close to 2. The fraction can be made arbitrarily close to 0 by taking $|x|$ large enough.

Consequently, $y = 0$ is an asymptote.

Finally, we let the degree of the numerator be greater than that of the denominator, as in

$$y = \frac{x^3 - 3x + 1}{3x^2 + 4x + 5} = \frac{x - (3/x) + (1/x^2)}{3 + (4/x) + (5/x^2)}$$

In this case $|y|$ will be larger than any chosen number when $|x|$ is sufficiently large.

We generalize this discussion by employing the equation

$$y = \frac{Ax^n + \text{(terms of lower degree)}}{Bx^m + \text{(terms of lower degree)}}$$

There are three possibilities here, depending on the relative values of m and n.

1. If $m = n$, $y = A/B$ is a horizontal asymptote.
2. If $m > n$, $y = 0$ is a horizontal asymptote.
3. If $m < n$, there is no horizontal asymptote.

Example 3. Draw the graph of the equation

$$y = \frac{(x + 3)(x - 1)}{(x + 1)(x - 2)}$$

Solution. The x intercepts are -3 and 1, and the y intercept is $\frac{3}{2}$. The lines $x = -1$ and $x = 2$ are vertical asymptotes. The numerator and denominator of the fraction are quadratic and the coefficient of x^2 in each is unity. Hence $y = 1$ is a horizontal asymptote. Since there are two vertical asymptotes, the graph consists of three separate parts. To aid in sketching, we examine the given equation to determine the signs of y to the right and left of each vertical asymptote and to the right and left of each x intercept. In this examination the notations $(+)$ and $(-)$ indicate the signs of the factors of the numerator and denominator for the specified values of x.

When $x < -3$, the signs are $\dfrac{(-)(-)}{(-)(-)}$, and hence $y > 0$.

When $-3 < x < -1$, the signs are $\dfrac{(+)(-)}{(-)(-)}$, and hence $y < 0$.

When $-1 < x < 1$, the signs are $\dfrac{(+)(-)}{(+)(-)}$, and hence $y > 0$.

When $1 < x < 2$, the signs are $\dfrac{(+)(+)}{(+)(-)}$, and hence $y < 0$.

When $x > 2$, the signs are $\dfrac{(+)(+)}{(+)(+)}$, and hence $y > 0$.

The asymptotes, the signs of y in the various intervals, and just a few plotted points in addition to the intercepts allow us to draw a good graph (Fig. 3).

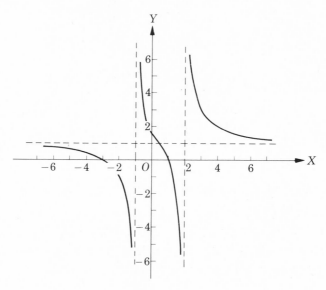

Figure 3

EXERCISES

Find the x intercepts of the graphs of the equations in Exercises 1 through 6. Determine the sign of y to the left of the smallest intercept, between consecutive intercepts, and to the right of the largest intercept. Then sketch the graph.

1. $y = x(x^2 - 4)$

2. $y = (x - 1)^2(x - 3)$

3. $y = x^2(4 - x)$

4. $y = x^4 - 16$

5. $y = (x + 2)^2(x^2 - 1)$

6. $y = (x^2 - 4)(x - 1)^2$

Discuss each of the following equations with regard to intercepts, symmetry, extent of curve, and vertical and horizontal asymptotes. Draw the asymptotes and sketch the graph.

7. $xy - x - 3 = 0$

8. $x^2y - 9y - 4 = 0$

9. $x^2y - 4y - 3x = 0$

10. $x^2y + y - x = 0$

11. $y = \dfrac{2x-3}{(x-1)^2}$

12. $y = \dfrac{x^2-4}{x^2+4}$

13. $y = \dfrac{x^2-9}{x^2-16}$

14. $y = \dfrac{x^2+4}{x^2-4}$

15. $y = \dfrac{(x+1)(x-2)}{x(x-4)}$

16. $y = \dfrac{(x-1)(x-2)}{(x+3)(x-1)}$

Find the equations of the horizontal asymptotes, when they exist, of the graphs of the following equations.

17. $y = \dfrac{x^{1.2}+3}{x-2}$

18. $y = \dfrac{x^3-x+1}{2x^2+x-3}$

19. $y = \dfrac{3x^4+5}{5x^4-7}$

20. $y = \dfrac{4x^5-x^4-2}{3x^6-x^5+3}$

21. $y = \dfrac{x^{2.4}+x^2-2}{x^{2.5}+x-10}$

22. $y = \dfrac{x-3}{x^{0.9}+3}$

2. IRRATIONAL EQUATIONS

In this section we consider equations in which y is equal to the square root of a polynomial in x or the square root of the quotient of two polynomials. The procedure for constructing the graphs is like that of the previous section.

Example 1. Draw the graph of the equation

$$y = \sqrt{x(x^2-16)}$$

Solution. The x intercepts are, 0, ± 4. The permissible values of x are those for which the radicand is not negative. The radicand is positive when $-4 < x < 0$ and also when $x > 4$. The radicand is negative when $x < -4$ and when $0 < x < 4$; hence these values must be excluded. This information and the few plotted points are sufficient for drawing the graph (Fig. 4).

x	-3	-2	-1	4.5	5
y (approx.)	4.6	4.9	3.9	4.4	6.7

Example 2. Draw the graph of the equation

$$y^2 = \dfrac{x^2-9}{x^2-16}$$

Solution. The graph is symmetric with respect to both axes. Hence we can draw the part in the first quadrant and readily finish the construction by the use of symmetry. The positive x intercept is 3, and the positive y intercept is $\frac{3}{4}$.

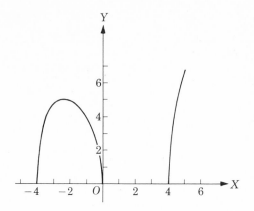

Figure 4

The line $x = 4$ is a vertical asymptote, and the line $y = 1$ is a horizontal asymptote. We observe that y^2 is positive when $0 \leq x < 3$ and when $x > 4$. But y^2 is negative when x has a value between 3 and 4; hence these values must be excluded. From the analysis of the given equation, we sketch the graph (Fig. 5). Somewhat greater accuracy can be had by plotting the points indicated in the accompanying table.

x	1	2	4.2	5	6	7
y (approx.)	0.7	0.6	2.3	1.3	1.2	1.1

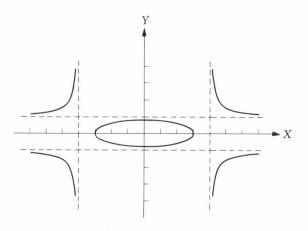

Figure 5

EXERCISES

Sketch the graph of each equation.

1. $y = \sqrt{x(x^2 - 4)}$

2. $y = \sqrt{x(x^2 - 9)}$

3. $y = \sqrt{\dfrac{x}{x^2 - 4}}$

4. $y = \sqrt{\dfrac{2x}{x^2 + 1}}$

5. $y^2 = \dfrac{4}{x + 3}$

6. $y^2 = \dfrac{4}{x^2 + 9}$

7. $y^2 = \dfrac{x^2 - 16}{x^2 - 9}$

8. $y^2 = \dfrac{x^2 - 9}{x^2 - 4}$

9. $y^2 = \dfrac{x(x - 1)}{x^2 - 4}$

10. $y^2 = \dfrac{x(x - 2)}{x - 1}$

11. $y^2 = \dfrac{x^2 - 4}{(x - 9)^2}$

12. $y^2 = \dfrac{x^2 - 16}{(x - 1)^2}$

3. SLANT ASYMPTOTES

When y is equal to the quotient of two polynomials in x where the degree of the numerator exceeds that of the denominator by unity, the graph will usually have a slant asymptote. The simplest case arises when the numerator is quadratic and the denominator is linear.

Example 1. Draw the graph of the equation

$$2xy - x^2 + 6x - 4y - 10 = 0$$

Solution. The $B^2 - 4AC$ test reveals that the graph is either a hyperbola or a degenerate conic (Chapter 4, Section 4). To determine which of these situations exists, we solve the equation for y. Thus, we get

$$y = \frac{x^2 - 6x + 10}{2x - 4}$$

and, by dividing,

$$y = \frac{1}{2}x - 2 + \frac{1}{x - 2}$$

For all real values of x, except $x = 2$, this equation yields real values for y. Hence the graph of the equation, and also the given equation, is a hyperbola, not a degenerate conic. Clearly, $x = 2$ is a vertical asymptote. As $|x|$ increases, the ordinate of the hyperbola gets closer and closer to the ordinate of the line $y = \frac{1}{2}x - 2$. And the difference of the two ordinates can be made arbitrarily near zero by taking $|x|$ large enough. Hence the line $y = \frac{1}{2}x - 2$ is an asymptote. The asymptote and just a few plotted points furnish a guide for drawing the graph (Fig. 6).

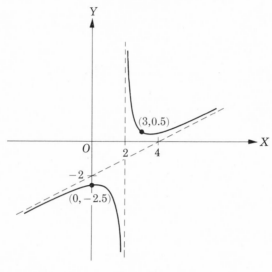

Figure 6

Example 2. Draw the graph of the equation

$$2x^2 y - x^3 - 8xy + 8x^2 - 20x + 8y + 14 = 0$$

Solution. On solving for y and performing a division, we obtain

$$y = \tfrac{1}{2}x - 2 + \frac{1}{(x-2)^2}$$

The line $x - 2 = 0$ is a vertical asymptote and the line $y = \frac{1}{2}x - 2$ is a slant asymptote. The fraction on the right side of the equation is positive for all permissible values of x. Consequently, the graph is above the slant asymptote, as shown in Fig. 7.

4. GRAPH OF AN EQUATION IN FACTORED FORM

Equations sometimes appear with one member equal to zero and the other member expressed as the product of factors in terms of x and y. When an

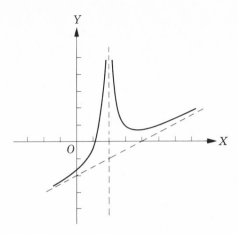

Figure 7

equation is in this form, its graph can be more simply obtained by first setting each of the factors equal to zero. If the coordinates of a point make one of the factors equal to zero, they make the product equal to zero and therefore satisfy the given equation. On the other hand, the coordinates of a point which make no factor equal to zero do not satisfy the equation. Hence the graph of the given equation consists of the graphs of the equations formed by setting each of the factors of the nonzero member equal to zero.

Example. The graph of the equation $(3x - y - 1)(y^2 - 9x) = 0$ consists of the line $3x - y - 1 = 0$ and the parabola $y^2 - 9x = 0$.

5. INTERSECTIONS OF GRAPHS

If the graphs of two equations in two variables have a point in common, then, from the definition of a graph, the coordinates of the point satisfy each equation separately. Hence the point of intersection gives a pair of real numbers which is a simultaneous solution of the equations. Conversely, if the two equations have a simultaneous real solution, then their graphs have the corresponding point in common. Thus simultaneous real solutions of two equations in two unknowns can be obtained graphically by reading the coordinates of their points of intersection. Because of the imperfections in the process, the results thus found are usually only approximate. If the graphs have no point of intersection, there is no real solution. In simple cases, the solutions, both real and imaginary, can be found by algebraic processes.

Example. Find the points of intersection of the graphs of

$$y = x^3 \qquad y = 2 - x$$

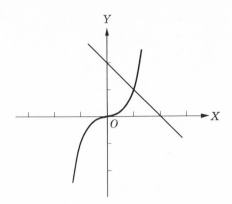

Figure 8

Solution. The graphs (Fig. 8) intersect in one point whose coordinates are $(1,1)$. Eliminating y between the equations yields

$$x^3 + x - 2 = 0 \qquad \text{or} \qquad (x-1)(x^2 + x + 2) = 0$$

Hence the roots of this equation are

$$x = 1 \qquad x = \frac{-1 + \sqrt{-7}}{2} \qquad x = \frac{-1 - \sqrt{-7}}{2}$$

The corresponding values of y are obtained from the linear equation. The solutions, real and imaginary, are

$$(1,\ 1) \qquad \left(\frac{-1 + \sqrt{-7}}{2}, \frac{5 - \sqrt{-7}}{2}\right) \qquad \left(\frac{-1 - \sqrt{-7}}{2}, \frac{5 + \sqrt{-7}}{2}\right)$$

The graphical method gives only the real solution.

EXERCISES

Draw the asymptotes and sketch the graph of each equation in Exercises 1 through 14.

1. $xy - x^2 + 2 = 0$

2. $xy + x^2 - 3 = 0$

3. $x^2 - xy + x + 1 = 0$

4. $x^2 + xy + 2x - 1 = 0$

5. $x^2 - xy + x - y + 2 = 0$

6. $x^2 - xy + 3x - 2y + 1 = 0$

7. $2x^2 + 2xy + 3x - 6y + 5 = 0$

8. $2x^2 - 2xy - x + y - 2 = 0$

9. $x^2 y - x^3 - 1 = 0$

10. $x^2 y - x^3 + 1 = 0$

11. $x^2y - x^3 - 4xy + 12x + 4y - 14 = 0$

12. $x^2y - x^3 - 4xy + 11x + 4y - 16 = 0$

13. $x^2y - x^3 - x^2 - x + y - 2 = 0$

14. $x^2y - x^3 - x^2 - x + y = 0$

Describe the graphs of the equations in Exercises 15 through 20.

15. $(x^2 + y^2)(x - y) = 0$

16. $(x^2 + y^2 + 1)(2x - 3y) = 0$

17. $xy(x + y - 2) = 0$

18. $2x^3 + 3xy^2 = 5x$

19. $x^3y + xy^3 = 4xy$

20. $x^2y - 9y^2 = 0$

Construct the graph of each pair of equations and estimate the coordinates of any points of intersection. Check by obtaining the solutions algebraically.

21. $x + 2y = 7$
    $3x - 2y = 5$

22. $x^2 + y^2 = 13$
    $3x - 2y = 0$

23. $x^2 - 4y = 0$
    $y^2 - 6x = 0$

24. $x^2 + y^2 = 16$
    $y^2 - 6x = 0$

25. $y = x^3 - 4x$
    $y = x + 4$

26. $x^2 - y^2 = 9$
    $x^2 + y^2 = 16$

27. $x^2 + 4y^2 = 25$
    $4x^2 - 7y^2 = 8$

28. $y = x^3 + 3x^2 - x - 3$
    $y = x + 5$

And the intersections of

Chapter 8

Space Coordinates and Surfaces

1. SPACE COORDINATES

In our study thus far, we have dealt with equations in two variables and have pictured equations in a plane coordinate system. When we introduce a third variable, a plane will not suffice for the illustration of an equation. For this purpose our coordinate system is extended to three dimensions.

Let OX, OY, and OZ be three mutually perpendicular lines (Fig. 1). These lines constitute the x axis, the y axis, and the z axis of a **three-dimensional rectangular coordinate system.** In this drawing, and others which we shall make, the y axis and the z axis are in the plane of the page. The x axis is to be

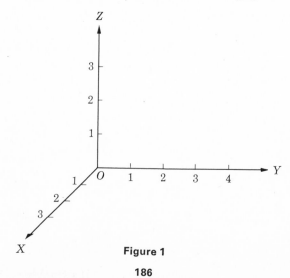

Figure 1

visualized as perpendicular to the page. The z axis may be regarded as vertical and the others as horizontal. The axes, in pairs, determine three mutually perpendicular planes called **coordinate planes.** The planes are designated the XOY plane, the XOZ plane, and the YOZ plane or, more simply, the xy plane, the xz plane, and the yz plane. The coordinate planes divide space into eight regions, called **octants.**

We next establish a number scale on each axis with the point O as the origin. The position of a point P in this coordinate system is determined by its distances from the coordinate planes. The distance of P from the yz plane is called the x **coordinate**, the distance from the xz plane the y **coordinate**, and the distance from the xy plane the z **coordinate**. The coordinates of a point are written in the form (x,y,z), in this order, x first, y second, and z third. To plot the point $(1.5,-1,2)$, for example, we go 1.5 units from the origin along the positive x axis, then 1 unit to the left parallel to the y axis, and finally 2 units upward parallel to the z axis. The signs of the coordinates determine the octant in which a point lies. Points whose coordinates are all positive are said to belong to the **first octant**; the other octants are not customarily assigned numbers. If a point is on a coordinate axis, two of its coordinates are zero.

In plotting points and drawing figures, we shall make unit distances on the y and z axes equal. A unit distance on the x axis will be represented by an actual length of about 0.7 of a unit. The x axis will be drawn at an angle of $135°$ with the y axis. This position of the x axis and the foreshortening in the x direction aid in visualizing space figures. Look at the cube and the plotted points in Fig. 2.

In our first application of the three-dimensional coordinate system, we consider the distance between two points of known coordinates. We took up this question for the two-dimensional case in Chapter 1, Section 4. Exactly the same plan will serve in our new system except for the additional axis. Thus, $P(x_1,y_1,z_1)$ and $Q(x_2,y_1,z_1)$ denote the endpoints of the line segment PQ parallel to the x axis. The distance from P to Q is $x_2 - x_1$. This distance is positive if $x_2 > x_1$ and negative if $x_2 < x_1$. In either case, using the absolute value symbols, $|PQ| = |x_2 - x_1|$. A similar situation applies to line segments parallel to the y axis and the z axis. With this understanding, we are ready to derive a formula for the distance between two points of known coordinates.

THEOREM 1   *Let $P_1(x_1,y_1,z_1)$ and $P_2(x_2,y_2,z_2)$ be the coordinates of two points in a three-dimensional coordinate system. Then the distance between P_1 and P_2 is given by*

$$|P_1 P_2| = \sqrt{(x_2 - x_1)^2 + (y_2 - y_1)^2 + (z_2 - z_1)^2}$$

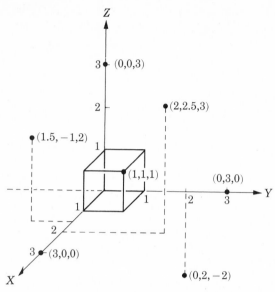

Figure 2

Proof. In Fig. 3 each edge of the rectangular parallelepiped (box-shaped figure) is parallel to a coordinate axis and each face is parallel to a coordinate plane. We indicate the coordinates of P_1, P_2, Q, and R, respectively, by

$$P_1(x_1, y_1, z_1) \qquad P_2(x_2, y_2, z_2) \qquad Q(x_2, y_1, z_1) \qquad R(x_2, y_2, z_1)$$

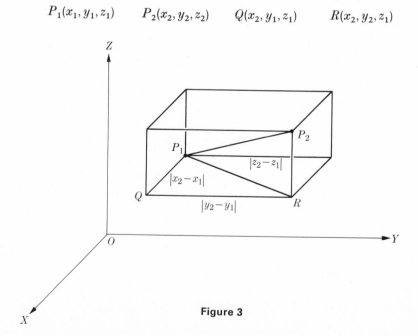

Figure 3

We observe that P_1QR is a right triangle with P_1R the hypotenuse, and P_1RP_2 is a right triangle with P_1P_2 the hypotenuse. Furthermore,

$$|P_1Q| = |x_2 - x_1| \qquad |QR| = |y_2 - y_1| \qquad |RP_2| = |z_2 - z_1|$$

Hence

$$|P_1P_2|^2 = |P_1R|^2 + |RP_2|^2$$
$$= |P_1Q|^2 + |QR|^2 + |RP_2|^2$$
$$= (x_2 - x_1)^2 + (y_2 - y_1)^2 + (z_2 - z_1)^2$$

or

$$\boxed{|P_1P_2| = \sqrt{(x_2 - x_1)^2 + (y_2 - y_1)^2 + (z_2 - z_1)^2}} \tag{1}$$

We have written the coordinates of four of the vertices of the parallelepiped. The student may write coordinates of the other four vertices.

Example. Find the distance between the points P_1 $(-4, 4, 1)$ and $P_2(-3, 5, -4)$.

Solution. Substituting in formula (1), we get

$$|P_1P_2| = \sqrt{(-3 + 4)^2 + (5 - 4)^2 + (-4 - 1)^2}$$
$$= \sqrt{1 + 1 + 25} = 3\sqrt{3}$$

2. THE GRAPH OF AN EQUATION

The graph of an equation in the three-dimensional system is defined exactly as in the two-dimensional system

DEFINITION 1 *The **graph** of an equation consists of the set of all points, and only those points, whose coordinates satisfy the equation.*

In studying graphs we shall use the idea of symmetry analogous to that discussed in Chapter 3, Section 3. For this purpose we state the following:

1. *If an equation is unchanged when x is replaced by −x, then the graph of the equation is symmetric with respect to the yz plane.*
2. *If an equation is unchanged when y is replaced by −y, then the graph of the equation is symmetric with respect to the xz plane.*
3. *If an equation is unchanged when z is replaced by −z, then the graph of the equation is symmetric with respect to the xy plane.*

In the two-dimensional system we found lines and curves as the graphs of equations. In three dimensions the graph of an equation is called a **surface**.

There are equations whose graphs, in three dimensions, are space curves (curves not lying in a plane). We are excluding space curves from consideration. We have observed, of course, that some two-dimensional equations have no graphs, and that others consist of one or more isolated points. Similarly, there are exceptional cases in a three-dimensional system. However, we shall be interested in equations whose graphs exist and are surfaces.

We shall begin our study of graphs by considering equations in one and two variables. As a further restriction, we shall use equations of only the first and second degrees. The graphs of equations of this class are comparatively easy to determine.

Example 1. To find the graph of the equation

$$y = 4$$

we observe that the equation is satisfied only by giving y the value 4. Since the equation does not contain x or z, no restrictions are placed on these variables; hence the graph consists of all points which have the y coordinate equal to 4. Clearly the graph is the plane parallel to the xz plane and 4 units to the right.

Example 2. Passing now to a linear equation in two variables, we choose for illustration the equation

$$2x + 3z = 6$$

In the xz plane this equation represents a line. Consider now a plane through this line and parallel to the y axis (Fig. 4). Any point $P(x,y,z)$ on this plane has a point on the line with the same x and z coordinates. Hence the coordinates of P satisfy the given equation. We conclude, therefore, that the plane is the graph of the equation.

The two examples illustrate the correctness of the following theorem.

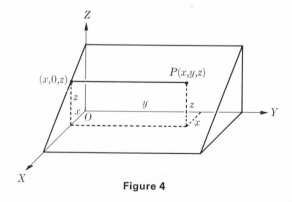

Figure 4

THEOREM 2 *The graph of a first-degree equation in one or two variables is a plane parallel to the axis of each missing variable.*

Example 3. Next, we discuss the equation

$$x^2 + (y - 2)^2 = 4$$

In the xy plane the graph of this equation is a circle of radius 2 with the center on the positive y axis 2 units from the origin (Fig. 5). Let $(x, y, 0)$ be the coordinates of any point of the circle. Then the point (x, y, z), where z is any real number,

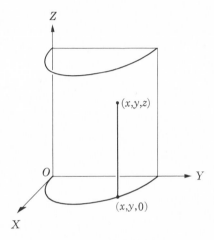

Figure 5

satisfies the equation. Thus we see that the graph of the given equation is a surface generated by a line which moves so that it keeps parallel to the z axis and intersects the circle. The surface therefore is a right circular cylinder which is symmetric with respect to the yz plane. In the figure, only the part of the surface in the first octant is indicated.

A surface generated by a line which moves so that it keeps parallel to a fixed line and intersects a fixed curve in a plane is called a **cylindrical surface** or **cylinder.** The curve is called the **directrix**, and the generating line in any position is called an **element** of the cylinder. In accordance with this definition, a plane is a special case of a cylinder with a straight line as the directrix. Hence the graph of each of the three equations which we have considered is a cylinder.

It is easy to generalize the preceding discussion to apply to equations in two variables, even without restriction to the degree, and establish the following theorem.

THEOREM 3   *The graph of an equation in two variables is a cylinder whose elements are parallel to the axis of the missing variable.*

3. THE GENERAL LINEAR EQUATION

In rectangular coordinates of two dimensions we found that a linear equation, in either one or two variables, represents a line. In our three-dimensional system we might, by analogy, surmise that linear equations in one, two, or three variables represent surfaces of the same type. The surmise is correct. At this point, however, we merely state the fact as a theorem and reserve the proof for the next chapter.

THEOREM 4   *The graph, in three dimensions, of the equation*

$$Ax + By + Cz + D = 0,$$

where the constants A, B, and C are not all zero, is a plane.

The lines in which a plane, or other surface, intersects the coordinate planes are called **traces.** The traces may be used to advantage in sketching the plane. For example, consider the plane

$$3x + 4y + 6z = 12$$

The x intercept of the plane, obtained by setting $y = 0$ and $z = 0$, is 4. Hence the plane passes through the point $(4,0,0)$. Similarly, the plane passes through the points $(0,3,0)$ and $(0,0,2)$. The traces of the given plane are the lines passing through these intercepts and forming the triangle drawn in Fig. 6. The triangle is an excellent aid in visualizing the position of the given plane.

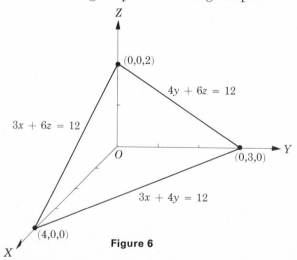

Figure 6

A plane parallel to a coordinate axis, and not passing through the origin, is easily sketched. For example, the trace of the plane $2y + 3z - 6 = 0$ in the yz plane passes through the points $(0,3,0)$ and $(0,0,2)$. The trace in the xy plane passes through $(0,3,0)$ and is parallel to the x axis, and the trace in the yz plane passes through $(0,0,2)$ and is parallel to the x axis.

EXERCISES

Plot the points in each of Exercises 1 through 8.

1. $(0,0,2)$ 2. $(0,2,0)$ 3. $(2,0,0)$

4. $(2,3,0)$ 5. $(3,2,4)$ 6. $(-2,0,4)$

7. $(-1,-1,-1)$ 8. $(2,1,-2)$ 9. $(4,-2,3)$

Find the distance between the points A and B in each of Exercises 10 through 13.

10. $A(-3,2,0)$, $B(6,-4,2)$ 11. $A(-2,1,2)$, $B(7,-5,4)$

12. $A(1,-5,2)$, $B(4,5,-2)$ 13. $A(3,-2,1)$, $B(-5,4,3)$

14. Draw a cube which has the origin and the point $(4,4,4)$ as opposite corners. Write the coordinates of the other corners if the cube has a face in each coordinate plane.

15. Draw the edges of a box which has four of its vertices at the points $(0,0,0)$, $(3,0,0)$, $(0,2,0)$, and $(0,0,2)$. What are the coordinates of the other vertices?

16. Draw the rectangular parallelepiped which has three of its faces in the coordinate planes and the line joining the points $(0,0,0)$ and $(4,5,3)$ as the ends of a diagonal. Write the coordinates of the remaining vertices.

Describe the surface corresponding to each of the following equations.

17. $x = 0$ 18. $y = 0$ 19. $z = 0$

20. $z = 5$ 21. $z = -5$ 22. $y = -3$

Describe the surface represented by each of the following equations. Make a sketch of the part of the surface in the first octant.

23. $3x + 2z = 6$ 24. $4y + 3z = 12$ 25. $5x + 3y = 15$

26. $x^2 + y^2 = 4$ 27. $y^2 + z^2 = 9$ 28. $4x^2 + 9y^2 = 36$

29. $(y - 3)^2 + z^2 = 9$ 30. $(x - 2)^2 + z^2 = 1$

31. $(x - 4)^2 + y^2 = 16$ 32. $2x + 3y + 4z = 12$

33. $2x + y + 2z = 4$ 34. $2x + 3y + 3z = 18$

4. SURFACE OF REVOLUTION

When a plane curve is revolved about a fixed line in the plane of the curve, a **surface of revolution** is said to be generated. The fixed line is called the **axis** of the surface. The path of each point of the curve is a circle with its center on the axis of the surface. Surfaces of revolution are used extensively in the applications of mathematics.

Example 1. Find the equation of the surface generated by revolving the ellipse

$$\frac{x^2}{a^2} + \frac{y^2}{b^2} = 1$$

about the x axis.

Solution. The surface is symmetric with respect to each coordinate plane. And, if $a > b$, the surface is shaped somewhat like a football. Let $P(x,y,z)$ be any point on the surface and let a plane perpendicular to the x axis pass through P (Fig. 7). The intersection of the plane and the surface is a circle. The center of

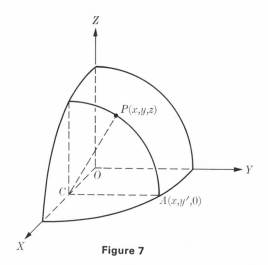

Figure 7

the circle is at $C(x,0,\ 0)$ and an intersection of the circle and the given ellipse is at $A(x,y',0)$. The segments CP and CA, being radii of a circle, have the same length. Hence $|CP|^2 = |CA|^2$. But

$$|CP|^2 = y^2 + z^2 \qquad \text{and} \qquad |CA|^2 = y'^2 = \frac{b^2}{a^2}(a^2 - x^2)$$

and, therefore

$$y^2 + z^2 = \frac{b^2}{a^2}(a^2 - x^2)$$

or

$$\frac{x^2}{a^2} + \frac{y^2}{b^2} + \frac{z^2}{b^2} = 1$$

This is the equation of the surface of revolution. Note that we would obtain this equation also by replacing y^2 by $y^2 + z^2$ in the given equation of the ellipse.

Example 2. A straight line, making a constant acute angle θ with the z axis, is rotated about the z axis. Find the equation of the surface thus generated.

Solution. Let $P(x, y, z)$ denote any point on the surface and let a plane perpendicular to the z axis pass through P (Fig. 8). The plane intersects the surface in a

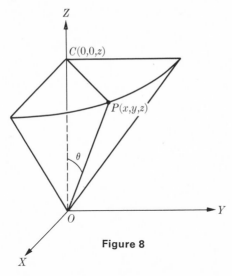

Figure 8

circle with center at $C(0, 0, z)$. The triangle OCP is a right triangle with C the vertex of the right angle. Consequently $|CP|/|OC| = \tan\theta$ and

$$|CP|^2 = |OC|^2 \tan^2\theta$$

But $|CP|^2 = x^2 + y^2$ and $|OC|^2 = z^2$, and therefore

$$x^2 + y^2 = z^2 \tan^2\theta$$

or

$$x^2 + y^2 - k^2 z^2 = 0$$

where $k = \tan\theta$. This is the equation of a right circular cone.

A sphere can be generated by revolving a circle (or just a semicircle) about a diameter. However, we shall not use the revolving method in finding the equation of a sphere.

Example 3. Find the equation of the set of points which are at a distance a from the point $C(h, k, \ell)$.

Solution. Using the distance formula (Section 1), we write immediately

$$(x - h)^2 + (y - k)^2 + (z - \ell)^2 = a^2 \qquad (2)$$

This is called the center-radius form of the equation of a spherical surface.

By performing the indicated squares in Eq. (2), we get an equation in the general form

$$x^2 + y^2 + Dx + Ey + Fz - H = 0 \qquad (3)$$

Conversely, we can reduce an equation of the form (3) to one of the form (2). We illustrate with the equation

$$x^2 + y^2 + z^2 - 2x + 6y + 8z + 17 = 0$$

Completing the squares in the x, y, and z terms gives

$$(x^2 - 2x + 1) + (y^2 + 6y + 9) + (z^2 + 8z + 16) = -17 + 1 + 9 + 16$$

or

$$(x - 1)^2 + (y + 3)^2 + (z + 4)^2 = 9$$

This equation, in center-radius form, represents a sphere with center at $(1, -3, -4)$ and radius 3.

5. SECOND-DEGREE EQUATIONS

The graph of a second-degree equation is called a **quadric surface**. It is not easy to determine the characteristics and location of a quadric surface corresponding to a general equation. To simplify the situation, we shall consider only second-degree terms which are squares of variables and not the product of two variables.

The main device in examining the graph of an equation consists of observing the intersections of the surface by the coordinate planes and planes parallel to them. The idea of symmetry, of course, may be used to advantage. To illustrate the method, we examine the equation

$$x^2 + y^2 = 4z$$

Replacing x by $-x$ and y by $-y$ does not alter the equation; hence the surface is symmetric to the yz and xz planes. Negative values must not be assigned to z and, consequently, no part of the surface is below the xy plane. Since $x = 0$, $y = 0$, and $z = 0$ satisfies the equation, the origin belongs to the surface. Sections (intersections) made by planes above and parallel to the xy plane are circles. This is evident if we substitute a positive value for z. The plane $z = 1$, for example, cuts the surface in the circle

$$x^2 + y^2 = 4$$

Circles of greater radii are obtained as the intersecting plane is taken farther and farther from the xy plane. We next substitute $y = 0$ in the given equation and get

$$x^2 = 4z$$

Hence the trace in the xz plane is a parabola. Similarly, the trace in the yz plane is the parabola $y^2 = 4z$.

We now have sufficient information to form a mental picture of the surface. As a matter of interest, though, we observe that sections parallel to the xz and yz planes are parabolas. When $x = 4$, for example, we find that the other coordinates must satisfy the equation

$$y^2 = 4(z - 4)$$

The coordinates of the vertex of this parabola are $(4,0,4)$. Fig. 9 shows a sketch of the surface.

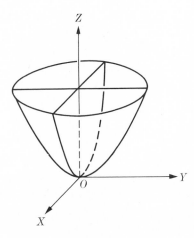

Figure 9

6. QUADRIC SURFACES

We shall now discuss a number of second-degree, or quadratic, equations which are said to be in standard forms. The study of these equations and their graphs, though presently of only geometric interest, furnish information and experience which will prove helpful in other mathematical situations, particularly in the calculus. Throughout this section we assume that a, b, and c stand for positive constants.

The ellipsoid

The surface represented by the equation

$$\frac{x^2}{a^2} + \frac{y^2}{b^2} + \frac{z^2}{c^2} = 1$$

is called an **ellipsoid**. We see at once that the surface is symmetric with respect to each coordinate plane. By setting one of the variables at a time equal to zero, we find the trace equations to be

$$\frac{x^2}{a^2} + \frac{y^2}{b^2} = 1 \qquad \frac{x^2}{a^2} + \frac{z^2}{c^2} = 1 \qquad \frac{y^2}{b^2} + \frac{z^2}{c^2} = 1$$

$$z = 0 \qquad\qquad y = 0 \qquad\qquad x = 0$$

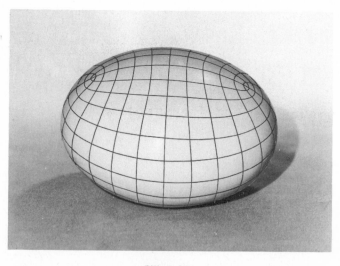

Ellipsoid

The traces are all ellipses. Next we assign to y a definite value $y = y_0$ such that $0 < y_0 < b$, and write the given equation in the form

$$\frac{x^2}{a^2} + \frac{z^2}{c^2} = 1 - \frac{y_0^2}{b^2}$$

This equation shows that sections made by planes parallel to the xz plane are ellipses. Furthermore, the elliptic sections decrease in size as the intersecting plane moves farther from the xz plane. When the moving plane reaches a distance b from the xz plane, the equation of the section becomes simply

$$\frac{x^2}{a^2} + \frac{z^2}{c^2} = 0$$

And the section, therefore, is the point $(0, b, 0)$. We see then that each of the planes $y = b$ and $y = -b$ contains one point of the ellipsoid, and all other points of the surface lie between these planes. Similarly, elliptic sections are obtained for values of x between $-a$ and a, and for z between $-c$ and c. So, by the method of sections, we get a clear mental picture of the ellipsoid (Fig. 10).

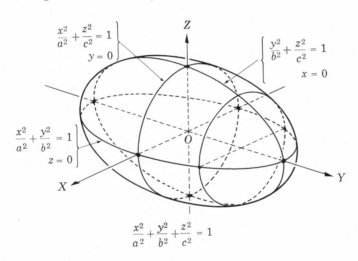

Figure 10

In our discussion, we have assumed that a, b, and c are any positive constants. If, however, two of the quantities are equal, the sections parallel to one of the coordinate planes are circles. Taking $a = c$, for example, and choosing an appropriate value y_0 for y, we have the equation

$$x^2 + z^2 = \frac{a^2}{b^2}(b^2 - y_0^2)$$

Thus we see that planes parallel to the xz plane cut the surface in circles. The ellipsoid in this instance could be generated by revolving the xy trace or the yz trace about the y axis. Finally, if $a = b = c$, the ellipsoid is a sphere.

The hyperboloid of one sheet

The surface represented by the equation

$$\frac{x^2}{a^2} + \frac{y^2}{b^2} - \frac{z^2}{c^2} = 1$$

is called a **hyperboloid of one sheet** (Fig. 11). The surface is symmetric with respect to each of the coordinate planes. Setting $z = 0$, we get the equation

$$\frac{x^2}{a^2} + \frac{y^2}{b^2} = 1$$

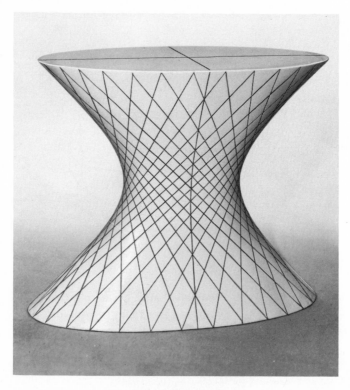

Hyperboloid of one sheet

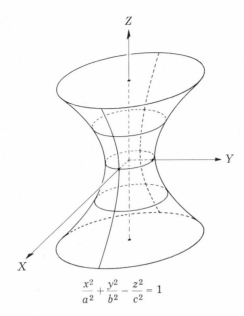

$$\frac{x^2}{a^2} + \frac{y^2}{b^2} - \frac{z^2}{c^2} = 1$$

Figure 11

Hence the xy trace is an ellipse. If we replace z in the given equation by a fixed value z_0, we obtain

$$\frac{x^2}{a^2} + \frac{y^2}{b^2} = 1 + \frac{z_0^2}{c^2}$$

This equation shows that sections parallel to the xy plane are ellipses and that the sections increase in size as the intersecting plane $z = z_0$ recedes from the origin. If $a = b$, the sections are circles, and the surface is a surface of revolution.

The traces in the xz and yz planes, respectively, are the hyperbolas

$$\frac{x^2}{a^2} - \frac{z^2}{c^2} = 1 \qquad \text{and} \qquad \frac{y^2}{b^2} - \frac{z^2}{c^2} = 1$$
$$y = 0 \qquad\qquad\qquad\qquad x = 0$$

The sections parallel to the xz and yz planes are likewise hyperbolas.

Each of the equations

$$\frac{x^2}{a^2} - \frac{y^2}{b^2} + \frac{z^2}{c^2} = 1 \qquad \text{and} \qquad -\frac{x^2}{a^2} + \frac{y^2}{b^2} + \frac{z^2}{c^2} = 1$$

represents a hyperboloid of one sheet. The first encloses the y axis and the second the x axis.

The hyperboloid of two sheets

The surface represented by the equation

$$\frac{x^2}{a^2} - \frac{y^2}{b^2} - \frac{z^2}{c^2} = 1$$

is called a **hyperboloid of two sheets**. The surface possesses symmetry with respect to each coordinate plane. By setting each variable in turn equal to zero, we get the equations

$$\frac{x^2}{a^2} - \frac{y^2}{b^2} = 1 \qquad \frac{x^2}{a^2} - \frac{z^2}{c^2} = 1 \qquad -\frac{y^2}{b^2} - \frac{z^2}{c^2} = 1$$
$$z = 0 \qquad\qquad y = 0 \qquad\qquad x = 0$$

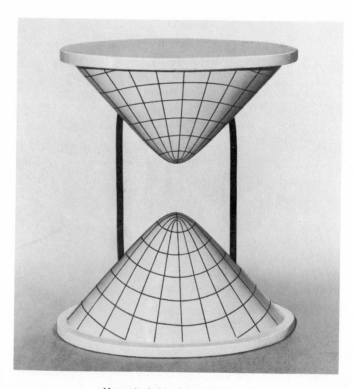

Hyperboloid of two sheets

These equations reveal that the xy and xz traces are hyperbolas, and that there is no trace in the yz plane. The sections made by the plane $x = x_0$ is given by the equation

$$\frac{y^2}{b^2} + \frac{z^2}{c^2} = \frac{x_0^2}{a^2} - 1$$

This equation represents a point or an ellipse according as the absolute value of x_0 is equal to or greater than a. Hence the graph of the given equation consists of two separate parts. Sections parallel to the xz and xy planes are hyperbolas. If $b = c$, the sections parallel to the yz plane are circles, and, in this case, the hyperboloid of two sheets is a surface of revolution.

Hyperboloids of two sheets are also represented by the equations

$$-\frac{x^2}{a^2} - \frac{y^2}{b^2} + \frac{z^2}{c^2} = 1 \qquad \text{and} \qquad -\frac{x^2}{a^2} + \frac{y^2}{b^2} - \frac{z^2}{c^2} = 1$$

The surface corresponding to this last equation is pictured in Fig. 12.

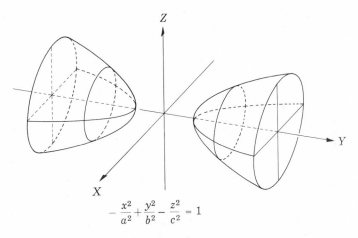

$$-\frac{x^2}{a^2} + \frac{y^2}{b^2} - \frac{z^2}{c^2} = 1$$

Figure 12

The elliptic paraboloid

The surface represented by the equation

$$\frac{x^2}{a^2} + \frac{y^2}{b^2} = cz$$

is called an **elliptic paraboloid.** The xy trace, obtained by setting $z = 0$, is the origin. Since we are assuming that a, b, and c are positive constants, the

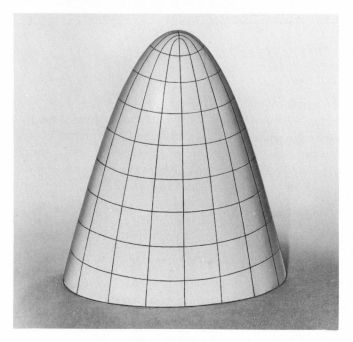

Elliptic paraboloid

surface, except for the origin, is above the xy plane. A plane parallel to the xy plane and cutting the surface makes an elliptic section which increases in size as the plane recedes from the origin. From this information, the surface (Fig. 13) can be readily visualized. If $a = b$, the sections parallel to the xy plane are circles. In this case the surface is obtainable by revolving either the xz trace or the yz trace about the z axis.

Elliptic paraboloids are also represented by the equations

$$\frac{x^2}{a^2} + \frac{z^2}{c^2} = by \qquad \text{and} \qquad \frac{y^2}{b^2} + \frac{z^2}{c^2} = ax$$

The hyperboloic parabloid

The surface represented by the equation

$$\frac{x^2}{a^2} - \frac{y^2}{b^2} = -cz$$

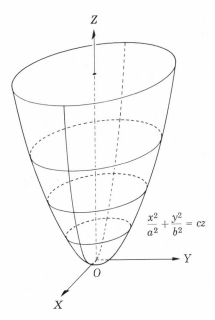

$$\frac{x^2}{a^2} + \frac{y^2}{b^2} = cz$$

Figure 13

is called the **hyperbolic paraboloid**. The surface is symmetric with respect to the yz and the xz planes. The xy trace is given by the pair of simultaneous equations

$$\frac{x^2}{a^2} - \frac{y^2}{b^2} = 0 \qquad \text{or} \qquad \left(\frac{x}{a} + \frac{y}{b}\right)\left(\frac{x}{a} - \frac{y}{b}\right) = 0$$
$$z = 0 \qquad\qquad\qquad\qquad\qquad\qquad z = 0$$

Hence the trace is a pair of lines intersecting at the origin. We represent the section of a plane parallel to the xy axis by the equation

$$\frac{x^2}{a^2} - \frac{y^2}{b^2} = -cz_0$$

This is a hyperbola with transverse axis parallel to the y axis when z_0 is positive and parallel to the x axis when z_0 is negative. Sections parallel to the xz plane and the yz plane are parabolas. The preceding analysis suggests that the hyperbolic paraboloid is a saddle-shaped surface. Further aid in visualizing the surface may be had from Fig. 14.

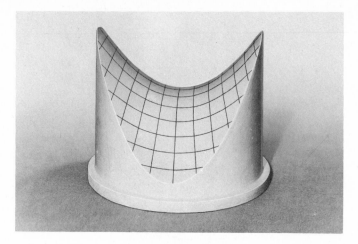

Hyperbolic Paraboloid

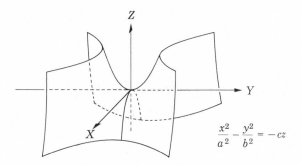

$$\frac{x^2}{a^2} - \frac{y^2}{b^2} = -cz$$

Figure 14

The elliptic cone

The surface represented by the equation

$$\frac{x^2}{a^2} + \frac{y^2}{b^2} = \frac{z^2}{c^2}$$

is called an **elliptic cone** (Fig. 15). Setting x, y, and z in turn equal to zero, we find the equations for the traces to be

$$\frac{x^2}{a^2} + \frac{y^2}{b^2} = 0 \qquad \frac{x^2}{a^2} = \frac{z^2}{c^2} \qquad \frac{y^2}{b^2} = \frac{z^2}{c^2}$$

$$z = 0 \qquad\qquad y = 0 \qquad\qquad x = 0$$

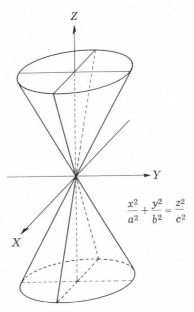

Figure 15

The equations reveal that the xy trace is the origin, and that each of the other traces is a pair of lines intersecting at the origin. Sections parallel to the xy plane are ellipses, and those parallel to the other coordinate planes are hyperbolas. If $a = b$, the cone is a right circular cone. Elliptic cones are also represented by the equations

$$\frac{x^2}{a^2} + \frac{z^2}{c^2} = \frac{y^2}{b^2} \qquad \text{and} \qquad \frac{y^2}{b^2} + \frac{z^2}{c^2} = \frac{x^2}{a^2}$$

EXERCISES

The following equations represent spheres. Find the coordinates of the center of the sphere and the radius.

1. $x^2 + y^2 + z^2 - 4y = 0$ 2. $x^2 + y^2 + z^2 - 6y - 2z = 0$

3. $x^2 + y^2 + z^2 - 4x + 8y - 6z - 7 = 0$ 4. $x^2 + y^2 + z^2 + 10x + 8y + 6z = 0$

5. Find the equation of the surface generated by revolving the ellipse

$$\frac{x^2}{a^2} + \frac{y^2}{b^2} = 1$$

about the y axis.

6. Find the equation of the surface generated by revolving the parabola $y^2 = 9x$ about the x axis.

Identify and sketch each quadric surface.

7. $\dfrac{x^2}{9} + \dfrac{y^2}{4} + \dfrac{z^2}{16} = 1$

8. $\dfrac{x^2}{9} + \dfrac{y^2}{9} + \dfrac{z^2}{4} = 1$

9. $\dfrac{x^2}{9} + \dfrac{y^2}{16} - \dfrac{z^2}{4} = 1$

10. $\dfrac{x^2}{4} - \dfrac{y^2}{4} + \dfrac{z^2}{9} = 1$

11. $-\dfrac{x^2}{16} + \dfrac{y^2}{9} - \dfrac{z^2}{4} = 1$

12. $\dfrac{x^2}{25} - \dfrac{y^2}{16} + \dfrac{z^2}{9} = 1$

13. $\dfrac{x^2}{9} + \dfrac{y^2}{4} = z$

14. $\dfrac{x^2}{16} + \dfrac{y^2}{9} = \dfrac{z^2}{4}$

Chapter 9

Vectors, Planes, and Lines

1. VECTORS

Many physical quantities possess the properties of magnitude and direction. A quantity of this kind is called a **vector quantity**. A force, for example, is characterized by its magnitude and direction of action. The force would not be completely specified by one of these properties without the other. The velocity of a moving body is determined by its speed (magnitude) and direction of motion. Acceleration and displacement are other examples of vector quantities.

Vectors are of great importance in physics and engineering. They are also used to much advantage in pure mathematics. The study of solid analytic geometry, in particular, is facilitated by the application of the vector concept. Our immediate objective in the introduction of vectors, however, is their use in dealing with planes and lines in space.

To obtain a geometric representation of a vector quantity, we employ a directed line segment (Chapter 1, Section 1) whose length and direction represent the magnitude and direction, respectively, of the vector quantity. In order to get a working basis for investigating problems involving vector quantities, we shall introduce certain definitions and establish a number of useful theorems.

DEFINITION 1 *A directed line segment, when used to represent a vector quantity, is called a* **vector**.

We shall denote a vector by a boldface letter or by giving its starting point and its ending point. Thus the vector (arrow) in Fig. 1 is drawn from O to P

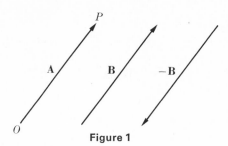

Figure 1

and we let $\overrightarrow{OP}$ or **A** indicate the vector. The point O is called the **foot** of the vector and the point P is called the **head**. The vectors **B** and $-$**B** in the figure have the same length as **A**. The vectors **A** and **B** have the same direction but $-$**B** is oppositely directed. The three vectors are related in accordance with the following definition.

DEFINITION 2 *Two vectors* **A** *and* **B** *are* **equal** (**A** = **B**) *if they have the same length and direction. The* **negative** *of a vector* **B**, *denoted by* $-$**B**, *is a vector having the same length as* **B** *but pointing in the opposite direction.*

In view of this definition, we note that vectors of the same length and different directions are not equal, and vectors of the same direction and different lengths are not equal.

2. OPERATIONS ON VECTORS

The operations of addition, subtraction, and multiplication of vectors are defined differently from the corresponding operations on real numbers. The new operations are designed so as to establish an appropriate theory for studying forces, velocities, and other physical concepts.

DEFINITION 3 *Let* **A** *and* **B** *denote vectors and draw from the head of* **A** *a vector equal to* **B**. *Then the* **sum** (**A** + **B**) *of* **A** *and* **B** *is the vector extending from the foot of* **A** *to the head of* **B**.

This definition is illustrated in Fig. 2. The sum of the vectors is called the **resultant** and each of the vectors forming the sum is called a **component**.

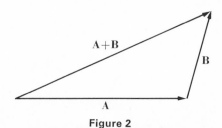

Figure 2

The triangle formed by the three vectors, **A**, **B**, and **A** + **B** is called a **vector triangle.** This method of adding vectors is used in physics, where it is shown, for example, that two forces applied at a point of a body have exactly the same effect as a single force equal to their resultant.

THEOREM 1   *Vector addition is commutative.*

Proof. We need to show that if **A** and **B** are any two vectors, then **A** + **B** = **B** + **A**. For this purpose we draw **A** and **B** from the point O (Fig. 3),

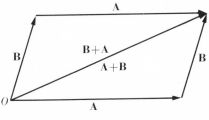

Figure 3

and then complete a parallelogram with the vectors forming adjacent sides. Since the opposite sides of a parallelogram are equal and parallel, we see that the foot of **B** in the lower part of the figure is at the head of **A**. Hence the sum **A** + **B** is along the diagonal extending from O. And from the upper part of the figure, we see that **B** + **A** is along the same diagonal. We conclude then that **A** + **B** = **B** + **A**, which means the vectors are commutative with respect to addition.

THEOREM 2   *Vectors obey the associative law of addition.*

Proof. Given three vectors, **A**, **B**, and **C**, we are to prove that

$$(\mathbf{A} + \mathbf{B}) + \mathbf{C} = \mathbf{A} + (\mathbf{B} + \mathbf{C})$$

From Fig. 4 we observe that $\overrightarrow{OE} = \mathbf{A} + \mathbf{B}$ and that **C** added to this sum yields

$$\overrightarrow{OF} = (\mathbf{A} + \mathbf{B}) + \mathbf{C}$$

Similarly, $\overrightarrow{DF} = \mathbf{B} + \mathbf{C}$ and adding this sum to **A** gives

$$\overrightarrow{OF} = \mathbf{A} + (\mathbf{B} + \mathbf{C})$$

Hence the sums $(\mathbf{A} + \mathbf{B}) + \mathbf{C}$ and $\mathbf{A} + (\mathbf{B} + \mathbf{C})$ have the same resultant, and consequently vector addition is associative.

DEFINITION 4   *A vector* **B** *subtracted from a vector* **A** *is equal to the sum of* **A** *and the negative of* **B**. *That is,*

$$\mathbf{A} - \mathbf{B} = \mathbf{A} + (-\mathbf{B})$$

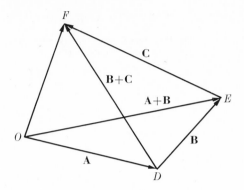

Figure 4

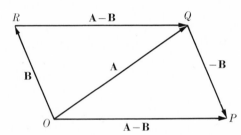

Figure 5

Referring to the parallelogram $OPQR$ in Fig. 5, we observe that the vector from O to P is equal to $\mathbf{A} - \mathbf{B}$, and the vector from R to Q is also equal to $\mathbf{A} - \mathbf{B}$. So, alternatively, if $\mathbf{A}$ and $\mathbf{B}$ are drawn from a common point, the vector from the head of $\mathbf{B}$ to the head of $\mathbf{A}$ is equal to $\mathbf{A} - \mathbf{B}$.

When numbers and vectors are involved in a problem or discussion, the numbers are sometimes called **scalars** to distinguish them from vectors.

DEFINITION 5   *The **product** of a scalar m and a vector* $\mathbf{A}$, *expressed by* $m\mathbf{A}$, *is a vector m times as long as* $\mathbf{A}$, *and has the direction of* $\mathbf{A}$ *if m is positive and the opposite direction if m is negative.*

This definition is illustrated in Fig. 6, where the scalar m has the values -2 and $\frac{3}{4}$.

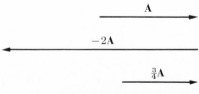

Figure 6

THEOREM 3   *If m and n are scalars and* **A** *and* **B** *are vectors, then*

$$(m + n)\mathbf{A} = m\mathbf{A} + n\mathbf{A} \qquad (1)$$

and

$$m(\mathbf{A} + \mathbf{B}) = m\mathbf{A} + m\mathbf{B} \qquad (2)$$

Proof. We leave the proof of Eq. (1) to the student. To establish Eq. (2), we note that **A, B,** and **A** + **B** form the sides of a triangle (Fig. 7). If each of these

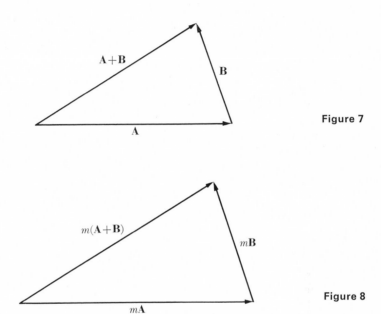

Figure 7

Figure 8

vectors is multiplied by a nonzero scalar m, the products $m\mathbf{A}$, $m\mathbf{B}$, and $m(\mathbf{A} + \mathbf{B})$ can be placed so as to form a triangle in which $m(\mathbf{A} + \mathbf{B}) = m\mathbf{A} + m\mathbf{B}$ (Fig. 8). Hence, as expressed by Eqs. (1) and (2), scalars and vectors obey the distributive law of multiplication.

Let us seek an interpretation of the difference of a vector and itself and the product of zero and a vector. That is, if **A** is a vector, what meaning should be given to **A** − **A** and $(0)\mathbf{A}$? For these quantities to be vectors, Definitions 4 and 5 require the length in each case to be equal to zero. To handle a situation of this kind, it is customary to enlarge the concept of a vector to include one of zero length, which is called the **zero vector.**

It is sometimes desirable to find two vectors whose sum is equal to a given vector. The given vector is then said to be **resolved** into two components. The components may be along any two directions in a plane containing the given vector. A graphical construction of the components may be obtained by forming a vector triangle of which the given vector is a side. The components are then along the other sides.

3. VECTORS IN A RECTANGULAR COORDINATE PLANE

In our consideration of vectors thus far we have not used a coordinate system. Many operations on vectors, however, can be carried out advantageously by the aid of a coordinate system. To begin our study of vectors in a coordinate plane, we introduce two special vectors each of unit length. One of the vectors, denoted by **i**, has the direction of the positive x axis; the other vector, denoted by **j**, has the direction of the positive y axis. Each of these vectors, as well as any vector of unit length, is said to be a **unit vector.**

Since vectors of the same length and same direction are equal (Definition 2), each of the vectors **i** and **j** may extend from any chosen point of the coordinate plane. But it is usually convenient to place them so that they extend from the origin (Fig. 9).

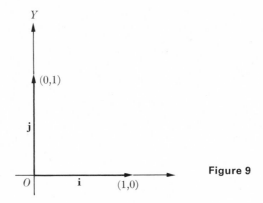

Figure 9

The product m**i** is a vector of length m units and has the direction of **i** if m is positive and the opposite direction if m is negative (Definition 5). A similar statement applies to the product m**j**. Using these facts, we shall point out that any vector can be expressed in terms of the unit vectors **i** and **j**. Let **V** be a vector from the origin to the point (a,b), as shown in Fig. 10. Clearly a**i** and b**j** are the horizontal and vertical components of the given vector, and therefore $\mathbf{V} = a\mathbf{i} + b\mathbf{j}$. The vector a**i** is called the x **component** of **V** and b**j** the y **component.**

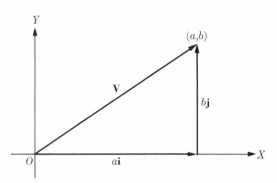

Figure 10

Continuing with the vector $\mathbf{V}$, we make the following observations. Since the length of the x component of $\mathbf{V}$ is a and the length of the y component is b, we may employ the Pythagorean theorem to find the length or magnitude of $\mathbf{V}$. Thus, denoting the length by $|\mathbf{V}|$, we have

$$|\mathbf{V}| = \sqrt{a^2 + b^2}$$

If $\mathbf{V}$ is divided by $|\mathbf{V}|$, the result is a unit vector with the same direction as that of $\mathbf{V}$. The length of $\mathbf{V} = 3\mathbf{i} - 4\mathbf{j}$, for example, is

$$|\mathbf{V}| = \sqrt{9 + 16} = 5$$

and

$$\frac{\mathbf{V}}{|\mathbf{V}|} = \frac{3\mathbf{i} - 4\mathbf{j}}{5} = \frac{3}{5}\mathbf{i} - \frac{4}{5}\mathbf{j}$$

is a unit vector having the same direction as $3\mathbf{i} - 4\mathbf{j}$.

THEOREM 4   *If the vectors V_1 and V_2, in terms of their x components and y components, are*

$$\mathbf{V}_1 = a_1\mathbf{i} + b_1\mathbf{j} \qquad and \qquad \mathbf{V}_2 = a_2\mathbf{i} + b_2\mathbf{j}$$

then

$$\mathbf{V}_1 + \mathbf{V}_2 = (a_1 + a_2)\mathbf{i} + (b_1 + b_2)\mathbf{j} \tag{3}$$

and

$$\mathbf{V}_1 - \mathbf{V}_2 = (a_1 - a_2)\mathbf{i} + (b_1 - b_2)\mathbf{j} \tag{4}$$

Proof. If the foot of $\mathbf{V}_1$ is at the origin, the head will be at the point (a_1, b_1). Then if the foot of $\mathbf{V}_2$ is at the head of $\mathbf{V}_1$, the head will be at the point $(a_1 + a_2, b_1 + b_2)$. And the vector from the origin to this point is expressed by $(a_1 + a_1)\mathbf{i} + (b_1 + b_2)\mathbf{j}$, which, by Definition 3, is equal to $\mathbf{V}_1 + \mathbf{V}_2$.

Although formula (3) follows readily from the definition of the sum of two vectors, it is worth noting that the formula can be established by use of Theorems 1, 2, and 3. Thus

$$\mathbf{V}_1 + \mathbf{V}_2 = (a_1\mathbf{i} + b_1\mathbf{j}) + (a_2\mathbf{i} + b_2\mathbf{j})$$
$$= a_1\mathbf{i} + a_2\mathbf{i} + b_1\mathbf{j} + b_2\mathbf{j}$$
$$= (a_1 + a_2)\mathbf{i} + (b_1 + b_2)\mathbf{j}$$

We leave the proof of formula (4) to the student.

Example 1. Vectors are drawn from the origin to the points $P(3, -2)$ and $Q(1,5)$. Indicating these vectors by $\overrightarrow{OP} = \mathbf{A}$ and $\overrightarrow{OQ} = \mathbf{B}$, find $\mathbf{A} + \mathbf{B}$ and $\mathbf{A} - \mathbf{B}$.

Solution. The given vectors, in terms of their x components and y components, are

$$\mathbf{A} = 3\mathbf{i} - 2\mathbf{j} \qquad \text{and} \qquad \mathbf{B} = \mathbf{i} + 5\mathbf{j}$$

Then, by the preceding theorem,

$$\mathbf{A} + \mathbf{B} = (3 + 1)\mathbf{i} + (-2 + 5)\mathbf{j}$$
$$= 4\mathbf{i} + 3\mathbf{j}$$

and

$$\mathbf{A} - \mathbf{B} = (3 - 1)\mathbf{i} + (-2 - 5)\mathbf{j}$$
$$= 2\mathbf{i} - 7\mathbf{j}$$

These two results are pictured in Fig. 11. The vector $\mathbf{A} - \mathbf{B}$, as we pointed out by use of Fig. 5, is equal to the vector extending from the head of $\mathbf{B}$ to the head of $\mathbf{A}$. The foot of $\mathbf{A} - \mathbf{B}$, in the figure, is not at the origin. But a vector equal to $\mathbf{A} - \mathbf{B}$ with its foot at the origin would have $(2,-7)$ as the coordinates of its head.

Example 2. Find the coordinates of the midpoint of the line segment joining the points $P(-2,4)$ and $Q(8,2)$.

Solution. We first find the vector from the origin to the midpoint of the line segment. This vector is equal to the vector from the origin to P plus half the vector from P to Q (Fig. 12). Indicating the vectors from the origin to P and Q by $\mathbf{A}$ and $\mathbf{B}$, respectively, we have

$$\mathbf{A} = -2\mathbf{i} + 4\mathbf{j}$$
$$\mathbf{B} = 8\mathbf{i} + 2\mathbf{j}$$
$$\mathbf{B} - \mathbf{A} = 10\mathbf{i} - 2\mathbf{j}$$

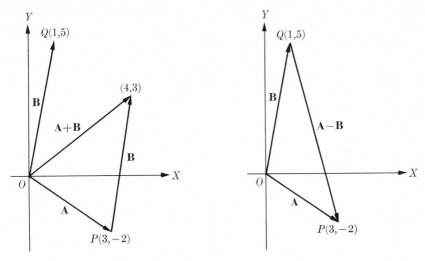

Figure 11

The desired vector **V** then is

$$\mathbf{V} = \mathbf{A} + \tfrac{1}{2}(\mathbf{B} - \mathbf{A})$$

$$= (-2\mathbf{i} + 4\mathbf{j}) + \tfrac{1}{2}(10\mathbf{i} - 2\mathbf{j})$$

$$= 3\mathbf{i} + 3\mathbf{j}$$

This result shows that the head of **V** is at the point $(3,3)$, and these are the coordinates of the midpoint of PQ.

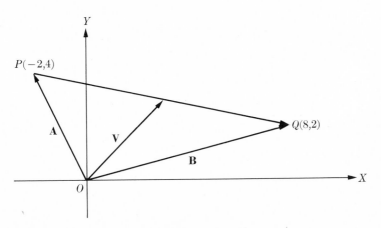

Figure 12

Example 3. Find the vectors from the origin to the trisection points of the line segment joining the points $P(1,3)$ and $Q(4,-3)$. Give the coordinates of the trisection points.

Solution. One of the required vectors is equal to the vector from the origin to P plus one-third of the vector from P to Q (Fig. 13). The other required

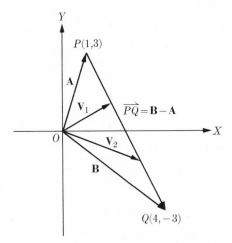

Figure 13

vector is equal to the vector from the origin to P plus two-thirds of the vector from P to Q. Denoting the vectors from the origin to P and Q by $\mathbf{A}$ and $\mathbf{B}$, respectively, we write

$$\mathbf{A} = \mathbf{i} + 3\mathbf{j}$$

$$\mathbf{B} = 4\mathbf{i} - 3\mathbf{j}$$

$$\mathbf{B} - \mathbf{A} = 3\mathbf{i} - 6\mathbf{j}$$

Hence one of the required vectors $\mathbf{V}_1$ is

$$\mathbf{V}_1 = \mathbf{A} + \tfrac{1}{3}(\mathbf{B} - \mathbf{A})$$

$$= (\mathbf{i} + 3\mathbf{j}) + \tfrac{1}{3}(3\mathbf{i} - 6\mathbf{j})$$

$$= 2\mathbf{i} + \mathbf{j}$$

The other vector $\mathbf{V}_2$ is

$$\mathbf{V}_2 = (\mathbf{i} + 3\mathbf{j}) + \tfrac{2}{3}(3\mathbf{i} - 6\mathbf{j})$$

$$= 3\mathbf{i} - \mathbf{j}$$

The vectors $2\mathbf{i} + \mathbf{j}$ and $3\mathbf{i} - \mathbf{j}$ tell us that the coordinates of the trisection points are $(2,1)$ and $(3,-1)$.

EXERCISES

In each of Exercises 1 through 4, let vectors extend from the origin to the points with the given coordinates. Then find the sum of the vectors. Also subtract the second vector from the first. Draw all vectors.

1. $P(2,3)$, $Q(-4,5)$ 2. $P(5,0)$, $Q(0,4)$

3. $P(3,-2)$, $Q(-1,-4)$ 4. $P(6,7)$, $Q(-5,-5)$

Determine a unit vector having the direction of the vector in each of Exercises 5 through 10.

5. $3\mathbf{i} + 4\mathbf{j}$ 6. $12\mathbf{i} - 5\mathbf{j}$ 7. $3\mathbf{i} - 12\mathbf{j}$

8. $\mathbf{i} + 2\mathbf{j}$ 9. $2\mathbf{i} - 3\mathbf{j}$ 10. $8\mathbf{i} + 15\mathbf{j}$

Use the method of Example 2 above to find the coordinates of the midpoint of the line segment joining P and Q in each of Exercises 11 through 14.

11. $P(3,6)$, $Q(5,-8)$ 12. $P(4,7)$, $Q(-2,-3)$

13. $P(-3,2)$, $Q(6,-4)$ 14. $P(-2,-3)$, $Q(5,4)$

15. Use vectors, as in the preceding exercises, to show that the coordinates of the midpoint of the line segment joining $P(a_1,b_1)$ and $Q(a_2 b_2)$ are

$$\left(\frac{a_1 + a_2}{2}, \frac{b_1 + b_2}{2} \right)$$

That is, the abscissa of the midpoint of a line segment is half the sum of the abscissas of the endpoints; the ordinate is half the sum of the ordinates of the endpoints.

Find the coordinates of the trisection points of the line segment joining P and Q in each of Exercises 16 through 19. Use the method of illustrative Example 3 above.

16. $P(-3,-3)$, $Q(6,3)$ 17. $P(-6,3)$, $Q(3,6)$

18. $P(-1,4)$, $Q(5,-2)$ 19. $P(-3,-5)$, $Q(2,7)$

20. By the use of vectors show that the coordinates of the trisection points of the line segment joining $P(a_1,b_1)$ and $Q(a_2,b_2)$ are

$$\left(\frac{2a_1 + a_2}{3}, \frac{2b_1 + b_2}{3} \right) \quad \text{and} \quad \left(\frac{a_1 + 2a_2}{3}, \frac{b_1 + 2b_2}{3} \right)$$

4. VECTORS IN SPACE

In the three-dimensional rectangular coordinate system, the unit vectors from the origin to the points $(1,0,0)$, $(0,1,0)$, and $(0,0,1)$ are denoted, respectively, by $\mathbf{i}$, $\mathbf{j}$, and $\mathbf{k}$. Any vector can be expressed in terms of these unit vectors. Thus the vector from the origin to the point $P(a, b, c)$ is given by

$$\overrightarrow{OP} = \mathbf{A} = a\mathbf{i} + b\mathbf{j} + c\mathbf{k}$$

The vectors $a\mathbf{i}$, $b\mathbf{j}$, and $c\mathbf{k}$ are the x, y, and z components of the vector $\mathbf{A}$. The length of $\mathbf{A}$ may be obtained by using the lengths of the sides of the vector right triangles OCP and ODC (Fig. 14). From the Pythagorean relation, we find

$$|\overrightarrow{OP}|^2 = |\overrightarrow{OC}|^2 + |\overrightarrow{CP}|^2$$

$$= |\overrightarrow{OD}|^2 + |\overrightarrow{DC}|^2 + |\overrightarrow{CP}|^2$$

$$= a^2 + b^2 + c^2$$

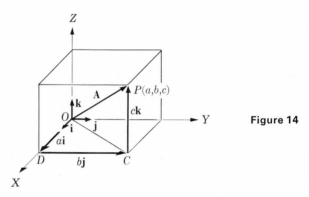

Figure 14

THEOREM 5   *If the vectors V_1 and V_2 in terms of their $x, y,$ and z components are*

$$\mathbf{V}_1 = a_1\mathbf{i} + b_1\mathbf{j} + c_1\mathbf{k} \qquad and \qquad \mathbf{V}_2 = a_2\mathbf{i} + b_2\mathbf{j} + c_2\mathbf{k}$$

then

$$\mathbf{V}_1 + \mathbf{V}_2 = (a_1 + a_2)\mathbf{i} + (b_1 + b_2)\mathbf{j} + (c_1 + c_2)\mathbf{k} \tag{5}$$

and

$$\mathbf{V}_1 - \mathbf{V}_2 = (a_1 - a_2)\mathbf{i} + (b_1 - b_2)\mathbf{j} + (c_1 - c_2)\mathbf{k} \tag{6}$$

Proof. If the foot of $\mathbf{V}_1$ is at the origin, the head will be at the point (a_1, b_1, c_1). Then if the foot of $\mathbf{V}_2$ is at the head of $\mathbf{V}_1$, the head of $\mathbf{V}_2$ will be at the point $[(a_1 + a_2), (b_1 + b_2), (c_1 + c_2)]$. And, by Definition 3, the vector from the origin to this point is equal to the sum of the two given vectors. Hence

$$\mathbf{V}_1 + \mathbf{V}_2 = (a_1 + a_2)\mathbf{i} + (b_1 + b_2)\mathbf{j} + (c_1 + c_2)\mathbf{k}$$

This establishes formula (5). We leave the proof of formula (6) to the student.

We remark that formulas (5) and (6) can also be obtained by use of Theorems 1, 2, and 3.

Example 1. Vectors extend from the origin to the points $P(3,-2,4)$ and $Q(1,5,-1)$. Indicating these vectors by $\overrightarrow{OP} = \mathbf{A}$ and $\overrightarrow{OQ} = \mathbf{B}$, find $\mathbf{A} + \mathbf{B}$ and $\mathbf{A} - \mathbf{B}$.

Solution. The given vectors in terms of their x, y, and z components are

$$\mathbf{A} = 3\mathbf{i} - 2\mathbf{j} + 4\mathbf{k} \qquad \text{and} \qquad \mathbf{B} = \mathbf{i} + 5\mathbf{j} - \mathbf{k}$$

Then, by the preceding theorem,

$$\mathbf{A} + \mathbf{B} = (3+1)\mathbf{i} + (-2+5)\mathbf{j} + (4-1)\mathbf{k} = 4\mathbf{i} + 3\mathbf{j} + 3\mathbf{k}$$

$$\mathbf{A} - \mathbf{B} = (3-1)\mathbf{i} + (-2-5)\mathbf{j} + (4+1)\mathbf{k} = 2\mathbf{i} - 7\mathbf{j} + 5\mathbf{k}$$

Example 2. Find the coordinates of the midpoint of the line segment joining the points $P(2,4,-1)$ and $Q(3,0,5)$.

Solution. We first find the vector from the origin to the midpoint of the line segment. This vector is equal to the vector from the origin plus half the vector from P to Q. Indicating the vectors from the origin to P and Q by $\mathbf{A}$ and $\mathbf{B}$, respectively, we have

$$\mathbf{A} = 2\mathbf{i} + 4\mathbf{j} - \mathbf{k}$$

$$\mathbf{B} = 3\mathbf{i} + 0\mathbf{j} + 5\mathbf{k}$$

$$\mathbf{B} - \mathbf{A} = \mathbf{i} - 4\mathbf{j} + 6\mathbf{k}$$

Since $\mathbf{B} - \mathbf{A}$ is the vector from the head of $\mathbf{A}$ to the head of $\mathbf{B}$, the desired vector $\mathbf{V}$ is

$$\mathbf{V} = \mathbf{A} + \tfrac{1}{2}(\mathbf{B} - \mathbf{A})$$

$$= (2\mathbf{i} + 4\mathbf{j} - \mathbf{k}) + \tfrac{1}{2}(\mathbf{i} - 4\mathbf{j} + 6\mathbf{k})$$

$$= \tfrac{5}{2}\mathbf{i} + 2\mathbf{j} + 2\mathbf{k}$$

From this result, we see that the head of $\mathbf{V}$ is at the point $(\tfrac{5}{2}, 2, 2)$, and these are the coordinates of the midpoint of PQ.

Example 3. Find the coordinates of the point which is $\tfrac{3}{4}$ of the way from $P(2,5,6)$ to $Q(6,-7,-2)$

Solution. Noticing that the vector from P to Q is $\overrightarrow{OQ} - \overrightarrow{OP}$, we desire to find the head of the vector $\overrightarrow{OP} + \tfrac{3}{4}(\overrightarrow{OQ} - \overrightarrow{OP})$. Thus we have

$$\overrightarrow{OP} = 2\mathbf{i} + 5\mathbf{j} + 6\mathbf{k}$$

$$\overrightarrow{OQ} = 6\mathbf{i} - 7\mathbf{j} - 2\mathbf{k}$$

$$\overrightarrow{OQ} - \overrightarrow{OP} = 4\mathbf{i} - 12\mathbf{j} - 8\mathbf{k}$$

Then the vector $\mathbf{V}$ from the origin to the point in question is

$$\mathbf{V} = 2\mathbf{i} + 5\mathbf{j} + 6\mathbf{k} + \tfrac{3}{4}(4\mathbf{i} - 12\mathbf{j} - 8\mathbf{k})$$

$$= 5\mathbf{i} - 4\mathbf{j} + 0\mathbf{k}$$

Hence $(5,-4,0)$ are the coordinates of the point $\tfrac{3}{4}$ of the way from P to Q.

EXERCISES

In each of Exercises 1 through 4, let vectors extend from the origin to the points P and Q. Then find the sum of the vectors. Also subtract the second vector from the first.

1. $P(3,2,4)$, $Q(5,-4,-1)$ 2. $P(0,5,4)$, $Q(4,0,7)$

3. $P(-2,3,4)$, $Q(-4,-1,0)$ 4. $P(6,7,5)$, $Q(5,-5,4)$

Determine a unit vector having the direction of the vector in each of Exercises 5 through 10.

5. $3\mathbf{i} + 2\mathbf{j} + 6\mathbf{k}$ 6. $4\mathbf{i} - \mathbf{j} - 8\mathbf{k}$ 7. $2\mathbf{i} + 2\mathbf{j} + \mathbf{k}$

8. $6\mathbf{i} - 3\mathbf{j} + 6\mathbf{k}$ 9. $2\mathbf{i} + \mathbf{j} + 3\mathbf{k}$ 10. $\mathbf{i} + \mathbf{j} + \mathbf{k}$

Find the vectors from the origin to the midpoint of the line segment joining the points P and Q in each of Exercises 11 through 14. What are the coordinates of the heads of these vectors?

11. $P(6,3,-8)$, $Q(4,5,2)$ 12. $P(4,7,5)$, $Q(2,-3,7)$

13. $P(5,6,7)$, $Q(3,4,9)$ 14. $P(8,-7,6)$, $Q(-3,4,6)$

Find the vectors from the origin to the trisection points of the line segment joining P and Q in Exercises 15 and 16. Give the coordinates of the heads of these vectors.

15. $P(1,-3,7)$, $Q(7,3,-2)$ 16. $P(4,3,-2)$, $Q(1,-3,4)$

Vectors extend from the origin to the points which divide the line segment joining P and Q into four equal parts in Exercises 17 and 18. Find the vectors and the coordinates of their heads.

17. $P(4,5,7)$, $Q(2,3,5)$ 18. $P(6,1,-1)$, $Q(-2,5,7)$

19. The line segment from $P(3,4,6)$ to $Q(-1,1,0)$ is produced by its own length through each end. Find the coordinates of the new ends.

5. THE SCALAR PRODUCT OF TWO VECTORS

So far we have not defined a product of two vectors. Actually two kinds of vector products are important in physics, engineering, and other fields. We shall introduce the simpler of the two products.

> DEFINITION 6   *The **scalar product** of two vectors A and B, denoted by*
> **A·B**, *is the product of their lengths times the cosine of the angle θ between*
> *them. That is,*
>
> $$\mathbf{A}\cdot\mathbf{B} = |\mathbf{A}||\mathbf{B}|\cos\theta$$

The name scalar is used because the product is a scalar quantity. The product is also called the **dot product**. It makes no difference whether the angle θ is positive or negative, since $\cos\theta = \cos(-\theta)$. However, we shall restrict θ to the interval $0° \leqslant \theta \leqslant 180°$. The angle is equal to $0°$ if A and B point in the same direction, and it is equal to $180°$ if they point oppositely.

Since $\cos 90° = 0$ and $\cos 0° = 1$, it is evident that the scalar product of two perpendicular vectors is zero, and the scalar product of two vectors in the same direction is the product of their lengths. The dot product of a vector on itself is the square of the length of the vector. That is,

$$\mathbf{A}\cdot\mathbf{A} = |\mathbf{A}|^2$$

In Fig. 15 the point M is the foot of the perpendicular to the vector **A** drawn from the tip of **B**. The vector from O to M is called the **vector projection** of **B** on **A**. The vector projection and **A** point in the same direction, since θ is an acute angle. If θ exceeds $90°$, then **A** and the vector from O to M point oppositely. The **scalar projection** of **B** on **A** is defined as $|\mathbf{B}|\cos\theta$; the sign of the scalar projection depends on $\cos\theta$. Using the idea of scalar projection of one vector on another, we can interpret the dot product geometrically as

$$\mathbf{A}\cdot\mathbf{B} = |\mathbf{A}||\mathbf{B}|\cos\theta$$

$$= \text{(length of } \mathbf{A}\text{) times (the scalar projection of } \mathbf{B} \text{ on } \mathbf{A}\text{)}$$

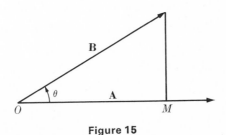

Figure 15

We could also say that the dot product of **A** and **B** is the length of **B** times the scalar projection of **A** on **B**.

By definition, **A·B** and **B·A** have exactly the same scalar factors and therefore

$$\mathbf{A \cdot B = B \cdot A} \tag{7}$$

This equation expresses the commutative property for the scalar multiplication of vectors. We next establish the distributive law.

THEOREM 6 *The scalar product of vectors is distributive. That is*

$$\mathbf{A \cdot (B + C) = A \cdot B + A \cdot C} \tag{8}$$

Proof. If we let b and c stand for the scalar projections of **B** and **C** on **A**, we see (Fig. 16) that the sum of the scalar projections of **B** and **C** on **A** is the same as the scalar projection of (**B** + **C**) on **A**. Hence

$$|\mathbf{A}|(b + c) = |\mathbf{A}|b + |\mathbf{A}|c$$

and

$$\mathbf{A \cdot (B + C) = A \cdot B + A \cdot C} \tag{9}$$

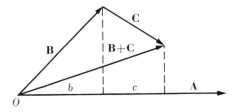

Figure 16

From Eqs. (7) and (8) we can deduce that the scalar product of sums of vectors may be carried out as in multiplying two algebraic expressions, each of more than one term. Thus, for example,

$$\mathbf{(A + B) \cdot (C + D) = A \cdot (C + D) + B \cdot (C + D)}$$
$$= \mathbf{A \cdot C + A \cdot D + B \cdot C + B \cdot D}$$

If m and n are scalars, then

$$(m\mathbf{A}) \cdot (n\mathbf{B}) = mn(\mathbf{A \cdot B}) \tag{10}$$

The equation is true if either m or n is equal to zero. But if m and n are both positive or both negative, then mn is positive, and consequently

$$(m\mathbf{A}) \cdot (n\mathbf{B}) = |m\mathbf{A}||n\mathbf{B}| \cos \theta$$
$$= mn(\mathbf{A \cdot B})$$

If m and n have opposite signs, the angle between $\mathbf{A}$ and $\mathbf{B}$ and the angle between $m\mathbf{A}$ and $n\mathbf{B}$ differ by $180°$. Then observing Fig. 17 and noting that mn is negative, we may write

$$(m\mathbf{A}) \cdot (n\mathbf{B}) = |m\mathbf{A}| \, |n\mathbf{B}| \cos(180° - \theta)$$

$$= -|m\mathbf{A}| \, |n\mathbf{B}| \cos\theta$$

$$= mn(\mathbf{A} \cdot \mathbf{B})$$

Hence Eq. (10) is true for all scalars m and n and all vectors $\mathbf{A}$ and $\mathbf{B}$.

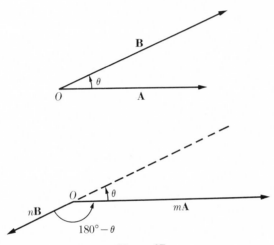

Figure 17

THEOREM 7   *If the vectors* $\mathbf{A}$ *and* $\mathbf{B}$ *are expressed in terms of the unit vectors* $\mathbf{i}, \mathbf{j},$ *and* $\mathbf{k}$ *by*

$$\mathbf{A} = a_1\mathbf{i} + b_1\mathbf{j} + c_1\mathbf{k}$$

$$\mathbf{B} = a_2\mathbf{i} + b_2\mathbf{j} + c_2\mathbf{k}$$

then

$$\mathbf{A} \cdot \mathbf{B} = a_1 a_2 + b_1 b_2 + c_1 c_2 \tag{11}$$

Proof. To obtain the dot product of $\mathbf{A}$ and $\mathbf{B}$, we determine first the dot products of the unit vectors $\mathbf{i}, \mathbf{j},$ and $\mathbf{k}$. Thus we have

$$\mathbf{i} \cdot \mathbf{i} = \mathbf{j} \cdot \mathbf{j} = \mathbf{k} \cdot \mathbf{k} = 1$$

$$\mathbf{i} \cdot \mathbf{j} = \mathbf{j} \cdot \mathbf{k} = \mathbf{k} \cdot \mathbf{i} = 0$$

Then we find

$$\mathbf{A} \cdot \mathbf{B} = (a_1 \mathbf{i} + b_1 \mathbf{j} + c_1 \mathbf{k}) \cdot (a_2 \mathbf{i} + b_2 \mathbf{j} + c_2 \mathbf{k})$$
$$= a_1 \mathbf{i} \cdot (a_2 \mathbf{i} + b_2 \mathbf{j} + c_2 \mathbf{k}) + b_1 \mathbf{j} \cdot (a_2 \mathbf{i} + b_2 \mathbf{j} + c_2 \mathbf{k})$$
$$+ c_1 \mathbf{k} \cdot (a_2 \mathbf{i} + b_2 \mathbf{j} + c_2 \mathbf{k})$$
$$= a_1 a_2 \mathbf{i} \cdot \mathbf{i} + 0 + 0 + 0 + b_1 b_2 \mathbf{j} \cdot \mathbf{j} + 0 + 0 + 0 + c_1 c_2 \mathbf{k} \cdot \mathbf{k}$$

Hence

$$\mathbf{A} \cdot \mathbf{B} = a_1 a_2 + b_1 b_2 + c_1 c_2$$

This equation shows that the dot product is obtained by the simple process of adding the products of the corresponding coefficients of $\mathbf{i}$, $\mathbf{j}$, and $\mathbf{k}$.

Example 1. Determine whether the vectors

$$\mathbf{A} = 3\mathbf{i} + 4\mathbf{j} - 8\mathbf{k}$$
$$\mathbf{B} = 4\mathbf{i} - 7\mathbf{j} - 2\mathbf{k}$$

are perpendicular.

Solution. The scalar product is

$$\mathbf{A} \cdot \mathbf{B} = (3)(4) + (4)(-7) + (-8)(-2) = 0$$

Since this product is zero, the vectors are perpendicular.

Example 2. Vectors are drawn from the origin to the points $P(6, -3, 2)$ and $Q(-2, 1, 2)$. Find the angle POQ.

Solution. Indicating $\overrightarrow{OP}$ by $\mathbf{A}$ and $\overrightarrow{OQ}$ by $\mathbf{B}$, we write

$$\mathbf{A} = 6\mathbf{i} - 3\mathbf{j} + 2\mathbf{k}$$
$$\mathbf{B} = -2\mathbf{i} + \mathbf{j} + 2\mathbf{k}$$

To find the angle, we substitute in both members of the equation

$$\mathbf{A} \cdot \mathbf{B} = |\mathbf{A}|\, |\mathbf{B}| \cos \theta$$

The product in the left member is $\mathbf{A} \cdot \mathbf{B} = -12 - 3 + 4 = -11$. The lengths of $\mathbf{A}$ and $\mathbf{B}$ are $|\mathbf{A}| = \sqrt{36 + 9 + 4} = 7$, $|\mathbf{B}| = \sqrt{4 + 1 + 4} = 3$. Hence

$$\cos \theta = \frac{\mathbf{A} \cdot \mathbf{B}}{|\mathbf{A}|\, |\mathbf{B}|} = \frac{-11}{21}$$

$$= -0.524$$

$$\theta = 122° \text{ nearest degree}$$

Example 3. Find the scalar projection and the vector projection of

$$\mathbf{B} = 2\mathbf{i} - 3\mathbf{j} - \mathbf{k} \qquad \text{on} \qquad \mathbf{A} = 3\mathbf{i} - 6\mathbf{j} + 2\mathbf{k}$$

Solution. The scalar projection of **B** on **A** is $|\mathbf{B}| \cos\theta$, where θ is the angle between the vectors. Using the equation

$$\mathbf{A} \cdot \mathbf{B} = |\mathbf{A}|\,|\mathbf{B}| \cos\theta$$

we have

$$|\mathbf{B}| \cos\theta = \frac{\mathbf{A} \cdot \mathbf{B}}{|\mathbf{A}|}$$

$$= \frac{6 + 18 - 2}{\sqrt{9 + 36 + 4}}$$

$$= \frac{22}{7}$$

The scalar projection of **B** on **A** is $\frac{22}{7}$. Since the scalar projection is positive, the vector projection is in the direction of **A**. The vector projection is, therefore, the product of the scalar projection and a unit vector in the direction of **A**. This unit vector is **A** divided by its length. Hence the vector projection of **B** on **A** is

$$\frac{22}{7} \cdot \frac{3\mathbf{i} - 6\mathbf{j} + 2\mathbf{k}}{7} = \frac{22}{49}(3\mathbf{i} - 6\mathbf{j} + 2\mathbf{k})$$

EXERCISES

Find the dot product of the vectors in Exercises 1 through 4. Find also the cosine of the angle between the vectors.

1. $\mathbf{A} = 4\mathbf{i} - \mathbf{j} + 8\mathbf{k}$
    $\mathbf{B} = 2\mathbf{i} + 2\mathbf{j} - \mathbf{k}$

2. $\mathbf{A} = 5\mathbf{i} - 3\mathbf{j} + 2\mathbf{k}$
    $\mathbf{B} = -\mathbf{i} + 7\mathbf{j} + 13\mathbf{k}$

3. $\mathbf{A} = 10\mathbf{i} + 2\mathbf{j} + 11\mathbf{k}$
    $\mathbf{B} = 4\mathbf{i} - 8\mathbf{j} - \mathbf{k}$

4. $\mathbf{A} = \mathbf{i} + \mathbf{j} + 2\mathbf{k}$
    $\mathbf{B} = 8\mathbf{i} + 4\mathbf{j} + \mathbf{k}$

In Exercises 5 and 6 find the scalar projection and the vector projection of **B** on **A**.

5. $\mathbf{A} = \mathbf{i} - \mathbf{j} - \mathbf{k}$
    $\mathbf{B} = 10\mathbf{i} - 11\mathbf{j} + 2\mathbf{k}$

6. $\mathbf{A} = 3\mathbf{i} + 3\mathbf{j} + \mathbf{k}$
    $\mathbf{B} = \mathbf{i} - 2\mathbf{j} - 2\mathbf{k}$

7. Find the angle which a diagonal of a cube makes with one of its edges.

8. From a vertex of a cube, a diagonal of a face and a diagonal of the cube are drawn. Find the angle thus formed.

The points in Exercises 9 and 10 are vertices of a triangle. In each, determine the vector from A to B and the vector from A to C. Find the angle between these vectors. Similarly, find the other interior angles of the triangle.

9. $A(3,4,2)$, $B(1,7,1)$, $C(-2,3,-5)$

10. $A(-2,-1,1)$, $B(1,0,-2)$, $C(0,-3,1)$

6. THE EQUATION OF A PLANE

We discovered in Chapter 8, Section 2 that a linear equation in one or two variables represents a plane. And we stated, without proof, that a linear equation in three variables also represents a plane (Theorem 3, Chapter 8). We now prove that a linear equation in one, two, or three variables represents a plane.

THEOREM 8   *Any plane in a three-dimensional rectangular coordinate system can be represented by a linear equation. Conversely, the graph of a linear equation is a plane.*

Proof. Suppose that a point $P_1(x_1, y_1, z_1)$ is in a given plane and that a nonzero vector

$$\mathbf{N} = A\mathbf{i} + B\mathbf{j} + C\mathbf{k}$$

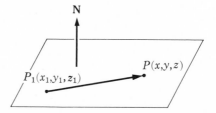

Figure 18

is perpendicular, or normal, to the plane (Fig. 18). A point $P(x,y,z)$ will lie in the given plane if, and only if, the vector

$$\overrightarrow{P_1 P} = (x - x_1)\mathbf{i} + (y - y_1)\mathbf{j} + (z - z_1)\mathbf{k}$$

is perpendicular to $\mathbf{N}$. Setting the scalar product of these vectors equal to zero, we obtain the equation

$$\mathbf{N} \cdot \overrightarrow{P_1 P} = 0$$

or

$$A(x - x_1) + B(y - y_1) + C(z - z_1) = 0 \tag{12}$$

This is the equation of the plane which passes through $P_1(x_1, y_1, z_1)$ and is perpendicular to the vector $\mathbf{N} = A\mathbf{i} + B\mathbf{j} + C\mathbf{k}$. Substituting D for the constant $-Ax_1 - By_1 - Cz_1$, we write the equation in the form

$$Ax + By + Cz + D = 0 \tag{13}$$

Conversely, any linear equation of the form (13) represents a plane. Starting with this equation, we can find a point $P_1(x_1, y_1, z_1)$ whose coordinates satisfy it. Then we have

$$Ax_1 + By_1 + Cz_1 + D = 0$$

This equation and Eq. (13) yield, by subtraction,

$$A(x - x_1) + B(y - y_1) + C(z - z_1) = 0$$

which is of the form (12). Hence Eq. (13) represents a plane perpendicular to the vector $\mathbf{N} = A\mathbf{i} + B\mathbf{j} + C\mathbf{k}$.

Example 1. Write the equation of the plane which contains the point $P_1(4, -3, 2)$ and is perpendicular to the vector $\mathbf{N} = 2\mathbf{i} - 3\mathbf{j} + 5\mathbf{k}$.

Solution. We use the coefficients of $\mathbf{i}$, $\mathbf{j}$, and $\mathbf{k}$ as the coefficients of x, y, and z and write the equation

$$2x - 3y + 5z + D = 0$$

For any value of D this equation represents a plane perpendicular to the given vector. The equation will be satisfied by the coordinates of the given point if

$$8 + 9 + 10 + D = 0 \qquad \text{or} \qquad D = -27$$

The required equation therefore is

$$2x - 3y + 5z - 27 = 0$$

Example 2. Find the equation of the plane determined by the points $P_1(1, 2, 6)$, $P_2(4, 4, 1)$ and $P_3(2, 3, 5)$.

Solution. A vector which is perpendicular to two sides of the triangle $P_1 P_2 P_3$ is normal to the plane of the triangle. To find such a vector, we write

$$\overrightarrow{P_1 P_2} = 3\mathbf{i} + 2\mathbf{j} - 5\mathbf{k}$$

$$\overrightarrow{P_1 P_3} = \mathbf{i} + \mathbf{j} - \mathbf{k}$$

$$\mathbf{N} = A\mathbf{i} + B\mathbf{j} + C\mathbf{k}$$

The coefficients A, B, and C are to be found so that $\mathbf{N}$ is perpendicular to each of the other vectors. Thus

$$\mathbf{N} \cdot \overrightarrow{P_1 P_2} = 3A + 2B - 5C = 0$$

$$\mathbf{N} \cdot \overrightarrow{P_1 P_3} = A + B - C = 0$$

These equations give $A = 3C$ and $B = -2C$. Choosing $C = 1$, we have $\mathbf{N} = 3\mathbf{i} - 2\mathbf{j} + \mathbf{k}$. The plane $3x - 2y + z + D = 0$ is normal to $\mathbf{N}$, and passes through the given points if $D = -5$. Hence

$$3x - 2y + z - 5 = 0$$

Example 4. Find the angle θ between the planes $4x - 8y - z + 5 = 0$ and $x + 2y - 2z + 3 = 0$.

Solution. The angle between two planes is equal to the angle between their normals. The vectors

$$\mathbf{N}_1 = \frac{4\mathbf{i} - 8\mathbf{j} - \mathbf{k}}{9} \qquad \text{and} \qquad \mathbf{N}_2 = \frac{\mathbf{i} + 2\mathbf{j} - 2\mathbf{k}}{3}$$

are unit vectors normal to the given planes. The dot product yields

$$\cos \theta = \mathbf{N}_1 \cdot \mathbf{N}_2 = -\tfrac{10}{27} \qquad \text{and} \qquad \theta = 112°$$

The planes intersect, making a pair of angles equal (approximately) to $112°$, and a second pair equal to $68°$. Choosing the smaller angle, we give the angle between the planes as $68°$.

THEOREM 9   *Let* $Ax + By + Cz + D = 0$ *be a plane and* $P_1(x_1, y_1, z_1)$ *a point not on the plane. Then the perpendicular distance* d *from the plane to* P_1 *is given by*

$$d = \frac{|Ax_1 + By_1 + Cz_1 + D|}{\sqrt{A^2 + B^2 + C^2}} \tag{14}$$

Proof. We let $P_2(x_2, y_2, z_2)$ be some point on the given plane and let $\mathbf{N} = \pm(A\mathbf{i} + B\mathbf{j} + C\mathbf{k})$, which is perpendicular to the plane, have its foot at P_2. The sign for $\mathbf{N}$ is to be chosen so that it will be on the same side of the plane as P_1, as pictured in Fig. 19. From the figure, we see that the desired distance d is equal to the scalar projection of $\overrightarrow{P_2 P_1}$ on $\mathbf{N}$. Observing that

$$\overrightarrow{P_2 P_1} = (x_1 - x_2)\mathbf{i} + (y_1 - y_2)\mathbf{j} + (z_1 - z_2)\mathbf{k}$$

we have

$$d = |\overrightarrow{P_2 P_1}| \cos\theta = \frac{\mathbf{N} \cdot \overrightarrow{P_2 P_1}}{|\mathbf{N}|}$$

$$= \frac{\pm(A\mathbf{i} + B\mathbf{j} + C\mathbf{k}) \cdot [(x_1 - x_2)\mathbf{i} + (y_1 - y_2)\mathbf{j} + (z_1 - z_2)\mathbf{k}]}{\sqrt{A^2 + B^2 + C^2}}$$

$$= \frac{\pm[A(x_1 - x_2) + B(y_1 - y_2) + C(z_1 - z_2)]}{\sqrt{A^2 + B^2 + C^2}}$$

$$= \frac{\pm(Ax_1 + By_1 + Cz_1 - Ax_2 - By_2 - Cz_2)}{\sqrt{A^2 + B^2 + C^2}}$$

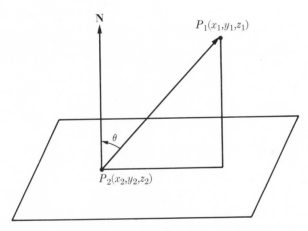

Figure 19

Since P_2 is on the plane, $-Ax_2 - By_2 - Cz_2 = D$. And to remove the ambiguity as to sign, we take the absolute value of the numerator and have

$$d = \frac{|Ax_1 + By_1 + Cz_1 + D|}{\sqrt{A^2 + B^2 + C^2}}$$

Example 5. Find the distance from the plane $2x + 3y - 6z - 2 = 0$ to the point $(4, -6, 1)$.

Solution. Substituting the proper values for A, B, C, D, x_1, y_1, and z_1 in formula (14), we get

$$d = \frac{|2(4) + 3(-6) + (-6)(1) - 2|}{\sqrt{2^2 + 3^2 + (-6)^2}}$$

$$= \frac{|8 - 18 - 6 - 2|}{\sqrt{4 + 9 + 36}}$$

$$= \frac{18}{7}$$

EXERCISES

Write the equation of the plane which satisfies the given conditions in Exercises 1 through 8.

1. Perpendicular to $N = 3i - 2j + 5k$ and passes through the point $(1, 1, 2)$.

2. Perpendicular to $N = 4i - j - k$ and passes through the origin.

3. Parallel to the plane $2x - 3y - 4z = 5$ and passes through $(1, 2, -3)$.

4. Perpendicular to the line segment joining $(4, 0, 6)$ and $(0, -8, 2)$ and passing through the midpoint of the segment.

5. Passes through the origin and is perpendicular to the line through $(2, -3, 4)$ and $(5, 6, 0)$.

6. Passes through the points $(0, 1, 2)$, $(2, 0, 3)$, $(4, 3, 0)$.

7. Passes through the points $(2, -2, -1)$, $(-3, 4, 1)$, $(4, 2, 3)$.

8. Passes through $(0, 0, 0)$, $(3, 0, 0)$, $(1, 1, 1)$.

Find the cosine of the acute angle between each pair of planes in Exercises 9 through 12.

9. $2x + y + z + 3 = 0$, $2x - 2y + z - 7 = 0$

10. $2x + y + 2z - 5 = 0$, $2x - 3y + 6z + 5 = 0$

11. $3x - 2y + z - 9 = 0$, $x - 3y - 9z + 4 = 0$

12. $x - 8y + 4z - 3 = 0$, $4x + 2y - 4z + 3 = 0$

13. Show that the planes

$$A_1 x + B_1 y + C_1 z + D_1 = 0 \quad \text{and} \quad A_2 x + B_2 y + C_2 z + D_2 = 0$$

are perpendicular if, and only if,

$$A_1 A_2 + B_1 B_2 + C_1 C_2 = 0$$

14. Determine the value of C so that the planes $2x - 6y + Cz = 5$ and $x - 3y + 2z = 4$ are perpendicular.

Find the distance from the given plane to the given point in Exercises 15 through 18.

15. $2x - y + 2z + 3 = 0$, $(0, 1, 3)$

16. $6x + 2y - 3z + 2 = 0$, $(2, -4, 3)$

17. $4x - 2y + z - 2 = 0$, $(-1, 1, 2)$

18. $3x - 4y - 5z = 0$, $(5, -1, 3)$

7. EQUATIONS OF A LINE

Let L be a line which passes through a given point $P_1(x_1, y_1, z_1)$ and is parallel to a given nonzero vector

$$\mathbf{V} = A\mathbf{i} + B\mathbf{j} + C\mathbf{k}$$

If $P(x, y, z)$ is a point on the line, then the vector $\overrightarrow{P_1 P}$ is parallel to $\mathbf{V}$ (Fig. 20). Conversely, if $\overrightarrow{P_1 P}$ is parallel to $\mathbf{V}$, the point P is on the line L. Hence P is on L if, and only if, there is a scalar t such that

$$\overrightarrow{P_1 P} = t\mathbf{V}$$

or

$$(x - x_1)\mathbf{i} + (y - y_1)\mathbf{j} + (z - z_1)\mathbf{k} = At\mathbf{i} + Bt\mathbf{j} + Ct\mathbf{k} \tag{15}$$

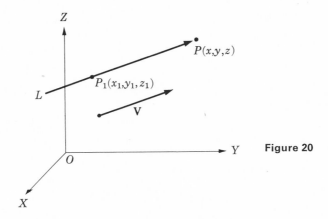

Figure 20

Equating corresponding coefficients of $\mathbf{i}$, $\mathbf{j}$, and $\mathbf{k}$, we obtain the equations

$$x - x_1 = At \qquad y - y_1 = Bt \qquad z - z_1 = Ct$$

or, transposing,

$$x = x_1 + At \qquad y = y_1 + Bt \qquad z = z_1 + Ct \tag{16}$$

When t is given any real value, Eqs. (16) determine the coordinates (x,y,z) of a point on the line L. Also there is a value of t corresponding to any point of the line. Equations (16) are called **parametric equations** of the line.

By solving each of the parametric equations for t and equating the equal values, we get

$$\frac{x - x_1}{A} = \frac{y - y_1}{B} = \frac{z - z_1}{C} \tag{17}$$

These are called the **symmetric equations** of the line.

The planes which contain a line and are perpendicular to the coordinate planes are called **projecting planes.**

Equations (17) represent three projecting planes. This becomes evident when we write the equations as

$$\frac{x - x_1}{A} = \frac{y - y_1}{B} \qquad \frac{x - x_1}{A} = \frac{z - z_1}{C} \qquad \frac{y - y_1}{B} = \frac{z - z_1}{C}$$

These equations, each in two variables, represent planes perpendicular, respectively, to the xy, xz, and yz planes. These equations represent a line, and hence the line is the intersection of the planes. Any two of the equations, of course, determine the line. We note also that any one of the equations can be obtained from the other two.

A line in space may be defined by two planes which pass through the line. Hence there are infinitely many ways of defining a line, since infinitely many planes pass through a line. However, it is usually convenient to deal with the projecting planes.

If a line is parallel to a coordinate plane, one of the quantities A, B, and C in Eqs. (17) is zero, and conversely. In this instance, one member of the equation would have zero in the denominator and could not be used. If, for example, $A = 0$ and B and C are not zero, then the line passing through $P_1(x_1,y_1,z_1)$ is parallel to the vector $\mathbf{V} = B\mathbf{j} + C\mathbf{k}$. Hence the line is parallel to the xy plane and, consequently, the plane $x = x_1$ contains the line. If two of A, B, and C are zero, say $A = B = 0$, then the line is parallel to the z axis. Thus the line is the intersection of the planes $x = x_1$ and $y = y_1$. So we see that when a denominator of a member of equations (17) is zero, the corresponding numerator equated to zero represents a plane through the line in question.

Example 1. Find symmetric equations which represent a line passing through the point $(2,-1,3)$ and parallel to the vector $\mathbf{V} = 2\mathbf{i} - 5\mathbf{j} + 6\mathbf{k}$. Also, find a set of parametric equations which represent the line.

Solution. The coefficients of $\mathbf{i}$, $\mathbf{j}$, and $\mathbf{k}$ of the given vector become the denominators in formula (17) and x_1, y_1, and z_1 are to be replaced, respectively, by $2, -1$, and 3. Thus we have

$$\frac{x-2}{2} = \frac{y+1}{-5} = \frac{z-3}{6}$$

The parametric equations, obtained by setting each member of these equations equal to a scalar t and solving for x, y, and z, are

$$x = 2 + 2t \quad y = -1 - 5t \quad z = 3 + 6t$$

The scalar t is a parameter which may be assigned any real value. Here we passed from the symmetric equations to the parametric equations. Conversely, of course, we can find the symmetric equations from the parametric equations.

Example 2. A line passes through the points $P_1(2, -4, 5)$ and $P_2(-1, 3, 1)$. Find symmetric equations of the line.

Solution. The vector from P_1 to P_2 is parallel to the line. Thus, we have

$$\overrightarrow{P_1 P_2} = -3\mathbf{i} + 7\mathbf{j} - 4\mathbf{k}$$

and, consequently, the equations

$$\frac{x-2}{-3} = \frac{y+4}{7} = \frac{z-5}{-4}$$

Example 3. Write the equations of the line passing through the points $P_1(2, 6, 4)$ and $P_2(3, -2, 4)$.

Solution. The vector from P_1 to P_2 is

$$\overrightarrow{P_1 P_2} = \mathbf{i} - 8\mathbf{j}$$

Hence the required line is parallel to the xy plane. The plane $z = 4$ contains the line. This plane is perpendicular to two of the coordinate planes. We use the first two members of Eqs. (17) to get another plane containing the line. Thus we have the defining equations

$$z = 4, \qquad \frac{x-3}{1} = \frac{y+2}{-8}$$

or

$$z = 4, \qquad 8x + y - 22 = 0$$

Note that we could not use the third member of the symmetric equations because its denominator would be zero. We did, however, set the numerator of that member equal to zero to obtain one of the planes.

Example 4. Find equations, in symmetric form, of the line of intersection of the planes

$$x + y - z - 7 = 0 \qquad \text{and} \qquad x + 5y + 5z + 5 = 0$$

Solution. We multiply the first equation by 5 and add to the second equation to eliminate z. We subtract the first equation from the second to eliminate x. This gives the equations

$$6x + 10y - 30 = 0 \qquad \text{and} \qquad 4y + 6z + 12 = 0$$

By solving each of these for y, we find

$$y = \frac{-3x + 15}{5} \qquad \text{and} \qquad y = \frac{-3z - 6}{2}$$

Therefore,

$$\frac{-3x + 15}{5} = y = \frac{-3z - 6}{2}$$

These symmetric equations can be converted to a simpler form if we divide each member by -3. This gives

$$\frac{x - 5}{5} = \frac{y}{-3} = \frac{z + 2}{2}$$

The symmetric equations can also be written by first finding the coordinates of two points on the line defined by the given equations. When $y = 0$ the equations become $x - z - 7 = 0$, $x + 5z + 5 = 0$. The solution of these equations is $x = 5$, $z = -2$. Hence the point $P_1(5, 0, -2)$ is on the line of intersection of the given planes. Similarly, we find that $P_2(0, 3, -4)$ is on the line. Then the vector

$$\overrightarrow{P_2 P_1} = 5\mathbf{i} - 3\mathbf{j} + 2\mathbf{k}$$

is parallel to the line whose equations we seek. Consequently, we obtain, as before, the symmetric equations

$$\frac{x - 5}{5} = \frac{y}{-3} = \frac{z + 2}{2}$$

8. DIRECTION ANGLES, DIRECTION COSINES, AND DIRECTION NUMBERS

We used the vector $\mathbf{V} = A\mathbf{i} + B\mathbf{j} + C\mathbf{k}$ in deriving the parametric equations of a line. The quantities A, B, and C, as we now point out, have a geometric significance with respect to the line.

DEFINITION 7   *The angles α, β, and γ which a directed line makes with the positive x, y, and z axes are called* **direction angles.** *The cosines of the direction angles are called* **direction** *cosines.*

The direction cosines of a line represented by equations of the form (16) or (17) may be found by the use of vectors. The vector

$$\mathbf{V} = A\mathbf{i} + B\mathbf{j} + C\mathbf{k}$$

is parallel to the line. And we may choose the positive direction of the line to be $\mathbf{V}$ or $-\mathbf{V}$. Choosing $\mathbf{V}$ as the positive direction, we take the scalar product of $\mathbf{V}$ and each of the unit vectors $\mathbf{i}$, $\mathbf{j}$, and $\mathbf{k}$. We note that the angle formed by two vectors which do not intersect is defined to be equal to the angle formed by two vectors which do intersect and are parallel to the given vectors and similarly directed.

Letting d stand for the length of $\mathbf{V}$, we have, by definition of the scalar product,

$$(A\mathbf{i} + B\mathbf{j} + C\mathbf{k})\cdot\mathbf{i} = |A\mathbf{i} + B\mathbf{j} + C\mathbf{k}|\,|\mathbf{i}|\cos\alpha$$

$$A = d\cos\alpha$$

Similarly, $B = d\cos\beta$ and $C = d\cos\gamma$. Hence

$$\cos\alpha = \frac{A}{d} \qquad \cos\beta = \frac{B}{d} \qquad \cos\gamma = \frac{C}{d}$$

The quantities A, B, and C are called **direction numbers.** Also the products of the direction cosines by any positive number are called direction numbers.

THEOREM 10   *The sum of the squares of the direction cosines of a line is equal to 1. That is,*

$$\cos^2\alpha + \cos^2\beta + \cos^2\gamma = 1$$

We leave the proof of this theorem to the student.

Example 1. The numbers 4, 1, and 8 are direction numbers of a line. Find the direction cosines of the line.

Solution. To get the direction cosines, we divide each direction number by $\sqrt{4^2 + 1^2 + 8^2} = 9$. This gives

$$\cos\alpha = \tfrac{4}{9} \qquad \cos\beta = \tfrac{1}{9} \qquad \cos\gamma = \tfrac{8}{9}$$

Example 2. A line makes an angle of $60°$ with the positive x axis and $45°$ with the positive y axis. What angle does the line make with the positive z axis?

Solution. Using Theorem 10, we have

$$\cos^2 60° + \cos^2 45° + \cos^2 \gamma = 1$$

$$\tfrac{1}{4} + \tfrac{1}{2} + \cos^2 \gamma = 1$$

Hence $\cos^2 \gamma = \tfrac{1}{4}$ and, therefore $\cos \gamma = \tfrac{1}{2}$ or $\cos \gamma = -\tfrac{1}{2}$. So $\gamma = 60°$ or $120°$.

The direction angles for a given line depend on the chosen positive direction of the line. We let α_1, β_1, and γ_1 denote the angles for one direction and α_2, β_2, and γ_2 for the opposite direction. Then

$$\alpha_2 = 180° - \alpha_1 \qquad \beta_2 = 180° - \beta_1 \qquad \gamma_2 = 180° - \gamma_1$$

These equations yield

$$\cos \alpha_2 = - \cos \alpha_1 \qquad \cos \beta_2 = - \cos \beta_1 \qquad \cos \gamma_2 = - \cos \gamma_1$$

So there are two possible sets of direction cosines for a line, one set being the negative of the other set.

Example 3. A line passes through $P_1(-4,9,5)$ and $P_2(2,12,3)$. Find the direction cosines if the line is directed from P_1 to P_2.

Solution. The vector from P_1 to P_2 is

$$\overrightarrow{P_1 P_2} = 6\mathbf{i} + 3\mathbf{j} - 2\mathbf{k}$$

Hence 6, 3, and -2 is a set of direction numbers. Since $\sqrt{6^2 + 3^2 + (-2)^2} = 7$, we have

$$\cos \alpha = \tfrac{6}{7} \qquad \cos \beta = \tfrac{3}{7} \qquad \cos \gamma = -\tfrac{2}{7}$$

Example 4. Assign a positive direction for the line represented by the equations

$$\frac{x-1}{4} = \frac{y+3}{-3} = \frac{z-5}{-2}$$

and find the direction cosines.

Solution. We can select $4, -3, -2$ as a set of direction numbers or the set $-4, 3, 2$. Selecting the second set and noting that $\sqrt{(-4)^2 + 3^2 + 2^2} = \sqrt{29}$, we have

$$\cos \alpha = \frac{-4}{\sqrt{29}} \qquad \cos \beta = \frac{3}{\sqrt{29}} \qquad \cos \gamma = \frac{2}{\sqrt{29}}$$

EXERCISES

In Exercises 1 through 8 find symmetric equations and parametric equations of the line which passes through P and is parallel to the given vector.

1. $P(4,-3,5)$; $-2\mathbf{i}+3\mathbf{j}+4\mathbf{k}$
2. $P(0,1,-2)$; $\mathbf{i}-\mathbf{j}+2\mathbf{k}$
3. $P(1,1,2)$; $2\mathbf{i}+3\mathbf{j}-\mathbf{k}$
4. $P(-2,-2,3)$; $5\mathbf{i}+4\mathbf{j}+\mathbf{k}$
5. $P(2,-1,1)$; $2\mathbf{i}+\mathbf{j}$
6. $P(3,3,3)$; $\mathbf{i}+\mathbf{j}$
7. $P(0,0,0)$; $\mathbf{j}$
8. $P(0,0,0)$; $\mathbf{k}$

Write symmetric equations for the line through P_1 and P_2 in Exercises 9 through 16.

9. $P_1(1,2,3)$, $P_2(-2,4,0)$
10. $P_1(0,0,0)$, $P_2(3,4,5)$
11. $P_1(1,0,2)$, $P_2(0,2,1)$
12. $P_1(2,4,0)$, $P_2(1,2,8)$
13. $P_1(2,5,4)$, $P_2(2,4,3)$
14. $P_1(0,4,3)$, $P_2(0,4,4)$
15. $P_1(-1,3,4)$, $P_2(3,3,4)$
16. $P_1(0,0,2)$, $P_2(0,0,4)$

Find a symmetric form for each pair of equations in Exercises 17 through 20.

17. $x-y-2z+1=0$
    $x-3y-3z+7=0$
18. $x+y-2z+8=0$
    $2x-y-2z+4=0$
19. $x+y+z-9=0$
    $2x+y-z+3=0$
20. $x+y-z+8=0$
    $2x-y+2z+6=0$

21. A line makes an angle of $45°$ with the positive x axis and $60°$ with the positive z axis. What angle does it make with the positive y axis?

22. A line makes an angle of $135°$ with the positive x axis and $60°$ with the positive y axis. What angle does it make with the positive z axis?

A line passes through P_1 and P_2 in Exercises 23 through 28. Find the direction cosines if the line is directed from P_1 to P_2.

23. $P_1(-3,4,-2)$, $P_2(1,5,6)$
24. $P_1(1,-1,4)$, $P_2(3,1,5)$
25. $P_1(-5,1,8)$, $P_2(5,3,-3)$
26. $P_1(3,1,-6)$, $P_2(0,5,6)$
27. $P_1(3,2,1)$, $P_2(4,5,6)$
28. $P_1(2,-3,4)$, $P_2(5,-6,7)$

29. The numbers 4, 1, 8 and 2, 2, 1 are direction numbers of two lines. Find the cosine of the acute angle θ between the lines.

30. The numbers 3, 4, 12 and 6, 3, 2 are direction numbers of two lines. Find the cosine of the acute angle θ between the lines.

By setting x, y, and z in turn equal to zero, find the coordinates of the points where the line cuts the coordinate planes.

31. $\dfrac{x-6}{2} = \dfrac{y+2}{1} = \dfrac{z+3}{3}$ 32. $\dfrac{x}{-2} = \dfrac{y-2}{1} = \dfrac{z-3}{1}$

33. $\dfrac{x-3}{3} = \dfrac{y}{-1} = \dfrac{z-4}{2}$ 34. $\dfrac{x-2}{1} = \dfrac{y+1}{2} = \dfrac{z-4}{3}$

9. THE VECTOR PRODUCT

We now introduce another kind of product of two vectors. Let **A** and **B** be nonparallel, nonzero vectors forming an angle θ, where $0 < \theta < 180°$ (Fig. 21).

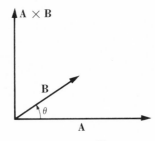

Figure 21

The **vector product,** or **cross product,** denoted by $\mathbf{A} \times \mathbf{B}$, is defined by the following statements:

1. $\mathbf{A} \times \mathbf{B}$ is a vector perpendicular to the plane determined by **A** and **B**.
2. $\mathbf{A} \times \mathbf{B}$ points in the direction a right-threaded screw would advance when its head is turned from **A** (the first vector) to **B** through the angle θ.
3. $|\mathbf{A} \times \mathbf{B}| = |\mathbf{A}| |\mathbf{B}| \sin \theta$.

If **A** and **B** are parallel ($\theta = 0°$ or $\theta = 180°$), the vector product is defined by statement 3. This makes the product equal to zero since $\sin \theta = 0$. The vector product is zero if either **A** or **B**, or both, is equal to zero.

From statement 2, we see that the interchange of the factors in vector multiplication reverses the direction of the product (Fig. 22). Hence

$$\mathbf{A} \times \mathbf{B} = -\mathbf{B} \times \mathbf{A},$$

and multiplication of this kind is not commutative.

The magnitude of the cross product has a simple geometric interpretation. The area of the parallelogram in Fig. 23 is $|\mathbf{A}|h$. But $h = |\mathbf{B}| \sin \theta$ and therefore

$$|\mathbf{A} \times \mathbf{B}| = |\mathbf{A}| |\mathbf{B}| \sin \theta = |\mathbf{A}|h$$

Hence the area of the parallelogram of which two adjacent sides are vectors is equal to the absolute value of the cross product of the vectors. One-half

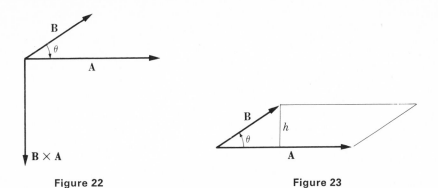

Figure 22   **Figure 23**

the absolute value of the cross product is, of course, equal to the area of the triangle determined by the vectors.

It can be shown that vector multiplication is distributive. (See Exercise 34, page 248.) The distributive property is expressed by the equation

$$\mathbf{A} \times (\mathbf{B} \times \mathbf{C}) = \mathbf{A} \times \mathbf{B} + \mathbf{A} \times \mathbf{C}$$

Suppose we apply the definition of vector multiplication to the unit vectors $\mathbf{i}, \mathbf{j}$, and $\mathbf{k}$ (Fig. 24). Clearly it follows that

$$\mathbf{i} \times \mathbf{j} = \mathbf{k} \quad \text{and} \quad \mathbf{j} \times \mathbf{i} = -\mathbf{k}$$

$$\mathbf{j} \times \mathbf{k} = \mathbf{i} \quad \text{and} \quad \mathbf{k} \times \mathbf{j} = -\mathbf{i}$$

$$\mathbf{k} \times \mathbf{i} = \mathbf{j} \quad \text{and} \quad \mathbf{i} \times \mathbf{k} = -\mathbf{j}$$

$$\mathbf{i} \times \mathbf{i} = \mathbf{j} \times \mathbf{j} = \mathbf{k} \times \mathbf{k} = 0$$

These equations and the distributive property of vector multiplication permit us to derive a convenient formula for the vector product when the vectors are expressed in terms of $\mathbf{i}, \mathbf{j}$, and $\mathbf{k}$. Thus, if

$$\mathbf{A} = a_1\mathbf{i} + b_1\mathbf{j} + c_1\mathbf{k} \quad \text{and} \quad \mathbf{B} = a_2\mathbf{i} + b_2\mathbf{j} + c_2\mathbf{k}$$

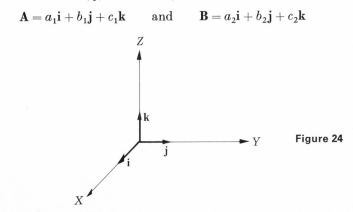

Figure 24

then

$$\mathbf{A} \times \mathbf{B} = (a_1\mathbf{i} + b_1\mathbf{j} + c_1\mathbf{k}) \times (a_2\mathbf{i} + b_2\mathbf{j} + c_2\mathbf{k})$$

$$= a_1 a_2 \mathbf{i} \times \mathbf{i} + a_1 b_2 \mathbf{i} \times \mathbf{j} + a_1 c_2 \mathbf{i} \times \mathbf{k}$$

$$+ a_2 b_1 \mathbf{j} \times \mathbf{i} + b_1 b_2 \mathbf{j} \times \mathbf{j} + b_1 c_2 \mathbf{j} \times \mathbf{k}$$

$$+ a_2 c_1 \mathbf{k} \times \mathbf{i} + b_2 c_1 \mathbf{k} \times \mathbf{j} + c_1 c_2 \mathbf{k} \times \mathbf{k}$$

$$= 0 + a_1 b_2 \mathbf{k} - a_1 c_2 \mathbf{j} - a_2 b_1 \mathbf{k} + 0$$

$$+ b_1 c_2 \mathbf{i} + a_2 c_1 \mathbf{j} - b_2 c_1 \mathbf{i} + 0$$

Hence

$$\mathbf{A} \times \mathbf{B} = (b_1 c_2 - b_2 c_1)\mathbf{i} + (a_2 c_1 - a_1 c_2)\mathbf{j} + (a_1 b_2 - a_2 b_1)\mathbf{k}$$

This equation provides a formula for the cross product; however, a more convenient form results when the right member of the equation is expressed as a determinant. Thus, alternatively, we discover that

$$\mathbf{A} \times \mathbf{B} = \begin{vmatrix} \mathbf{i} & \mathbf{j} & \mathbf{k} \\ a_1 & b_1 & c_1 \\ a_2 & b_2 & c_2 \end{vmatrix}$$

Example 1. The vectors $\mathbf{A}$ and $\mathbf{B}$ make an angle of $30°$. If the length of $\mathbf{A}$ is 5 and the length of $\mathbf{B}$ is 8, find the area of the parallelogram of which $\mathbf{A}$ and $\mathbf{B}$ are consecutive sides.

Solution. The area of the parallelogram is equal to the absolute value of $\mathbf{A} \times \mathbf{B}$. From the definition of a cross product,

$$|\mathbf{A} \times \mathbf{B}| = |\mathbf{A}| \, |\mathbf{B}| \sin \theta = (5)(8) \sin 30° = 20$$

The area of the parallelogram is 20 square units.

Example 2. The points $A(1,0,-1)$, $B(3,-1,-5)$, and $C(4,2,0)$ are vertices of a triangle. Find the area of the triangle.

Solution. The vectors $\overrightarrow{AB}$ and $\overrightarrow{AC}$ form two sides of the triangle. The magnitude of the cross product of the vectors is equal to the area of the parallelogram

of which the vectors are adjacent sides. The area of the triangle, however, is
one-half the area of the parallelogram. Thus we find

$$\vec{AB} = 2\mathbf{i} - \mathbf{j} + 4\mathbf{k} \qquad \vec{AC} = 3\mathbf{i} + 2\mathbf{j} + \mathbf{k}$$

and

$$\vec{AB} \times \vec{AC} = \begin{vmatrix} \mathbf{i} & \mathbf{j} & \mathbf{k} \\ 2 & -1 & -4 \\ 3 & 2 & 1 \end{vmatrix} = 7\mathbf{i} - 14\mathbf{j} + 7\mathbf{k}$$

The magnitude of this vector is $\sqrt{49 + 196 + 49} = 7\sqrt{6}$. Therefore the area
of the triangle is $\frac{7}{2}\sqrt{6}$.

Example 3. The points $A(2,-1,3)$, $B(4,2,5)$, and $C(-1,-1,6)$ determine a
plane. Find the distance from the plane to the point $D(5,4,8)$.

Solution. The vectors $\vec{AB}$ and $\vec{AC}$ determine a plane as indicated in Fig. 25.
The cross product $\vec{AB} \times \vec{AC}$ is perpendicular to the plane, and $\vec{AD}$ is the

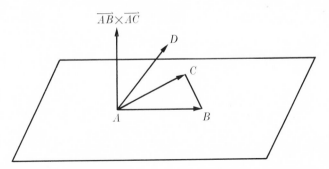

Figure 25

vector from A to the given point D. The scalar product of $\vec{AD}$ and a unit vector
perpendicular to the plane yields the required distance d. Accordingly, we
write

$$\vec{AD} = 3\mathbf{i} + 5\mathbf{j} + 5\mathbf{k} \qquad \vec{AB} = 2\mathbf{i} + 2\mathbf{j} + 2\mathbf{k}, \qquad \vec{AC} = -3\mathbf{i} + 3\mathbf{k}$$

$$\vec{AB} \times \vec{AC} = \begin{vmatrix} \mathbf{i} & \mathbf{j} & \mathbf{k} \\ 2 & 3 & 2 \\ -3 & 0 & 3 \end{vmatrix} = 9\mathbf{i} - 12\mathbf{j} + 9\mathbf{k}$$

The vector $9\mathbf{i} - 12\mathbf{j} + 9\mathbf{k}$ is perpendicular to $\vec{AB}$ and $\vec{AC}$ and therefore is
perpendicular to the plane of A, B, and C. We divide this vector by its length,

$3\sqrt{34}$, to obtain a unit vector. Finally, the distance d from the plane to D is given by

$$d = (3\mathbf{i} + 5\mathbf{j} + 5\mathbf{k}) \cdot \frac{3\mathbf{i} - 4\mathbf{j} + 3\mathbf{k}}{\sqrt{34}} = \frac{2\sqrt{34}}{17}$$

Example 4. Find equations, in symmetric form, of the line of intersection of the planes $2x - y + 4z = 3$ and $3x + y + z = 7$.

Solution. We may write the desired equations immediately if we know the coordinates of any point of the line and any vector parallel to the line. By setting $z = 0$ in the equations of the planes and solving for x and y, we find that $(2, 1, 0)$ is a point of the line. Normals to the planes are given by the vectors

$$\mathbf{N_1} = 2\mathbf{i} - \mathbf{j} + 4\mathbf{k} \qquad \text{and} \qquad \mathbf{N_2} = 3\mathbf{i} + \mathbf{j} + \mathbf{k}$$

The cross product of the vectors is parallel to the line of intersection of the planes. We find

$$\mathbf{N_1} \times \mathbf{N_2} = \begin{vmatrix} \mathbf{i} & \mathbf{j} & \mathbf{k} \\ 2 & -1 & 4 \\ 3 & 1 & 1 \end{vmatrix} = -5\mathbf{i} + 10\mathbf{j} + 5\mathbf{k}$$

We divide this vector by 5 and write the equation of the line as

$$\frac{x - 2}{-1} = \frac{y - 1}{2} = \frac{z}{1}$$

We note that this is the same kind of problem as that of Example 4, Section 7.

10. THE DISTANCE FROM A LINE TO A POINT

Let L (Fig. 26) be a line passing through the point $P_0(x_0, y_0, z_0)$ and parallel to a unit vector $\mathbf{u}$. Then let $P_1(x_1, y_1, z_1)$ be any point not on the line and let $\mathbf{V}$ be the vector $\overrightarrow{P_0P_1}$. To find the perpendicular distance from line L to P_1, we take the vector product of $\mathbf{V}$ and $\mathbf{u}$. Thus

$$d = |\mathbf{V}| \sin \theta = |\mathbf{u}| |\mathbf{V}| \sin \theta = |\mathbf{u} \times \mathbf{V}|.$$

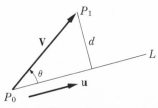

Figure 26

Example. Find the distance from the line passing through $P_2(3,0,6)$ and $P_3(5,-2,7)$ to the point $P_1(8,1,-3)$.

Solution. The vector $\overrightarrow{P_3P_2} = -2\mathbf{i} + 2\mathbf{j} - \mathbf{k}$, and $\mathbf{u} = \frac{1}{3}(-2\mathbf{i} + 2\mathbf{j} - \mathbf{k})$. The vector $\overrightarrow{P_2P_1} = \mathbf{V} = 5\mathbf{i} + \mathbf{j} - 9\mathbf{k}$. Hence

$$\mathbf{u} \times \mathbf{V} = \frac{1}{3} \begin{vmatrix} \mathbf{i} & \mathbf{j} & \mathbf{k} \\ -2 & 2 & -1 \\ 5 & 1 & -9 \end{vmatrix} = \frac{1}{3}(-17\mathbf{i} - 23\mathbf{j} - 14\mathbf{k})$$

and

$$d = |\mathbf{u} \times \mathbf{V}| = \frac{13}{3}\sqrt{6}$$

In Chapter 2, Section 3, we derived a formula for the distance from a line to a point when both are in the xy plane. Now we give a much shorter derivation by employing vectors.

The equation $Ax + By + C = 0$, in three space, represents a plane parallel to the z axis (Chapter 8, Theorem 2); the vector $A\mathbf{i} + B\mathbf{j}$ is perpendicular to the plane (Chapter 9, Section 6). The intersection of the plane and the xy plane is a line, and the x and y coordinates of all points of the intersection satisfy the equation $Ax + By + C = 0$. Restricting ourselves to the xy plane, we have a line and a vector perpendicular to the line. Now let $P_1(x_1, y_1)$ be any point in the xy plane (Fig. 27). If $B \neq 0$, the line cuts the y axis at

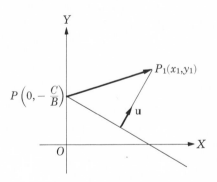

Figure 27

$P(0, -C/B)$. The scalar product of the vector $\overrightarrow{PP_1}$ and a unit vector $\mathbf{u}$, perpendicular to the line, gives the distance d. Since

$$\overrightarrow{PP_1} = x\mathbf{i} + \left(y_1 + \frac{C}{B}\right)\mathbf{j} \quad \text{and} \quad \mathbf{u} = \frac{A\mathbf{i} + B\mathbf{j}}{\pm\sqrt{A^2 + B^2}}$$

we have

$$d = \left[x_1\mathbf{i} + \left(y_1 + \frac{C}{B} \right)\mathbf{j} \right] \cdot \frac{A\mathbf{i} + B\mathbf{j}}{\pm\sqrt{A^2 + B^2}} = \frac{Ax_1 + By_1 + C}{\pm\sqrt{A^2 + B^2}}$$

Thus a scalar product yields at once the desired formula. The ambiguity as to sign is discussed in Chapter 2, Section 3.

EXERCISES

In Exercises 1 through 6 find the cross product, $\mathbf{A} \times \mathbf{B}$, of the vectors and a unit vector perpendicular to the two given vectors.

1. $\mathbf{A} = 3\mathbf{i} - 4\mathbf{j} - 2\mathbf{k}$,
   $\mathbf{B} = \mathbf{i} - 2\mathbf{j} - 2\mathbf{k}$

2. $\mathbf{A} = 6\mathbf{i} - 3\mathbf{j} + 14\mathbf{k}$,
   $\mathbf{B} = 3\mathbf{i} - 2\mathbf{j} + 3\mathbf{k}$

3. $\mathbf{A} = \mathbf{i} + \mathbf{j} + \mathbf{k}$,
   $\mathbf{B} = \mathbf{i} - \mathbf{j} - \mathbf{k}$

4. $\mathbf{A} = 2\mathbf{i} - \mathbf{k}$,
   $\mathbf{B} = \mathbf{j} + 2\mathbf{k}$

5. $\mathbf{A} = 4\mathbf{i} - 3\mathbf{j}$,
   $\mathbf{B} = 3\mathbf{i} + 4\mathbf{j}$

6. $\mathbf{A} = \mathbf{i} - 2\mathbf{j} + 3\mathbf{k}$,
   $\mathbf{B} = 4\mathbf{i} + 5\mathbf{j} - 6\mathbf{k}$

Compute the area of the parallelogram described in Exercises 7 through 10.

7. $\mathbf{A} = 3\mathbf{i} + 2\mathbf{j}$ and $\mathbf{B} = \mathbf{i} - \mathbf{j}$ are adjacent sides.

8. $\mathbf{A} = 4\mathbf{i} - \mathbf{j} + \mathbf{k}$ and $\mathbf{B} = 3\mathbf{i} + \mathbf{j} + \mathbf{k}$ are adjacent sides.

9. The points $A(4,1,0)$, $B(1,2,1)$, $C(0,0,6)$, and $D(3,-1,5)$ are vertices of the parallelogram $ABCD$.

10. The points $A(1,-1,1)$, $B(3,3,1)$, $C(4,-1,4)$, and $D(2,-5,4)$ are vertices of the parallelogram $ABCD$.

Find the area of the triangle described in Exercises 11 through 14.

11. The vertices are $A(-1,3,4)$, $B(1,2,5)$, and $C(2,-3,1)$.

12. The vertices are $A(1,0,4)$, $B(3,-3,0)$, and $C(0,1,2)$.

13. The vectors $\mathbf{A} = 2\mathbf{i} - 3\mathbf{j}$ and $\mathbf{B} = \mathbf{i} - \mathbf{j}$, drawn from the origin, are two sides.

14. The tips of the vectors $\mathbf{A} = 4\mathbf{i} + \mathbf{j}$, $\mathbf{B} = \mathbf{i} + 2\mathbf{j} + \mathbf{k}$, and $\mathbf{C} = 3\mathbf{i} - \mathbf{j} + 5\mathbf{k}$, drawn from the origin, are the vertices.

15. Find the area of the triangle in the xy plane whose vertices are the points (x_1, y_1), (x_2, y_2) and (x_3, y_3). Use a vector product and compare your result with the formula in Exercise 34, Chapter 1, page 11.

16. Write the equation of the plane which passes through the points $(2,-4,3)$, $(-3,5,1)$, and $(4,0,6)$.

17. Write the equation of the plane which contains the points $(2,1,0)$, $(3,0,2)$, and $(0,4,3)$.

18. Find the distance from the point $(6,7,8)$ to the plane of the points $(-1,3,0)$, $(2,2,1)$, $(1,1,3)$.

19. Find the distance from the point $(-4,-5,-3)$ to the plane passing through the points $(4,0,0)$, $(0,6,0)$, and $(0,0,7)$.

20. Find the equations of the line passing through the point $(3,1,-2)$ and parallel to each of the planes $x - y + z = 4$ and $3x + y - z = 5$.

21. Find the equations of the line which passes through the origin and is parallel to the line of intersection of the planes $2x - y - z = 2$ and $4x + 2y - 4z = 1$.

22. Find a parametric representation of the line of intersection of the two planes of Exercise 20 and also of Exercise 21.

23. A plane passes through the point $(1,1,1)$ and is perpendicular to each of planes $2x + 2y + z = 3$ and $3x - y - 2z = 5$. Find its equation.

24. A plane passes through the point $(0,0,0)$ and is perpendicular to the line of intersection of the planes $5x - 4y + 3z = 2$ and $x + 2y - 3z = 4$. Find the equation of the plane.

25. A line passes through the points $(1,3,1)$ and $(3,4,-1)$. Find the distance from the line to the point $(4,4,4)$.

26. A line passes through $(3,2,1)$ and is parallel to the vector $2\mathbf{i} + \mathbf{j} - 2\mathbf{k}$. Find the distance from the line to $(-3,-1,3)$.

27. Find the distance from the line $x/2 = y/3 = z/1$ to the point $(3,4,1)$.

28. Find the distance from the line $(x - 2)/2 = y/2 = (z - 1)/1$ to the point $(0,0,0)$.

29. Let L_1 and L_2 (Fig. 28) be lines which are not parallel and do not intersect. Let $\overrightarrow{P_1P_2}$ be a vector extending from any point on L_1 to any point on L_2, $\mathbf{V}_1$ a vector parallel to L_1, and $\mathbf{V}_2$ parallel to L_2. Then $\mathbf{V}_1 \times \mathbf{V}_2$ is perpen-

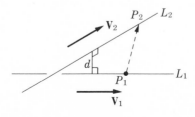

Figure 28

dicular to both lines. Show that the perpendicular distance between L_1 and L_2 is

$$d = |\overrightarrow{P_1P_2} \cdot \mathbf{u}|$$

where $\mathbf{u}$ is a unit vector parallel to $\mathbf{V}_1 \times \mathbf{V}_2$.

Find the perpendicular distance between the two lines in Exercises 30 through 33.

30. L_1 passes through $P_1(2,3,1)$ in the direction of $\mathbf{V}_1 = \mathbf{i} + 2\mathbf{j} - 3\mathbf{k}$, and L_2 passes through $P_2(4,2,0)$ in the direction of $\mathbf{V}_2 = 3\mathbf{i} - \mathbf{j} + \mathbf{k}$.

31. L_1 passes through $(2,1,-1)$ and $(-1,3,2)$. L_2 passes through $(4,0,5)$ and $(3,4,0)$.

32. The equations of L_1 are $(x+2)/3 = (y+3)/2 = z/2$. The equations of L_2 are $(x-1)/2 = (y+4)/3 = (z-2)/4$.

33. L_1 passes through $(4,1,0)$ and $(0,0,6)$. L_2 passes through $(1,2,1)$ and $(3,-1,5)$. Do the lines intersect?

34. In Fig. 29, the vector $\mathbf{E}$ is the projection of $\mathbf{B}$ on a plane perpendicular to $\mathbf{A}$. Vector $\mathbf{E}$ is rotated through $90°$, as indicated, and then multiplied by $|\mathbf{A}|$ to yield $\mathbf{A} \times \mathbf{B}$. Tell why these operations lead to $\mathbf{A} \times \mathbf{B}$.

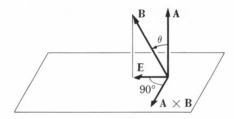

Figure 29

Next, let $\mathbf{B}$, $\mathbf{C}$, and $\mathbf{B} + \mathbf{C}$ form a triangle. Project this triangle on the plane perpendicular to $\mathbf{A}$, thus forming another vector triangle. Rotate the new triangle through $90°$ and multiply each side by $|\mathbf{A}|$. Observe that the sides of the final triangle are $\mathbf{A} \times \mathbf{B}$, $\mathbf{A} \times \mathbf{C}$, and $\mathbf{A} \times (\mathbf{B} + \mathbf{C})$, and that

$$\mathbf{A} \times (\mathbf{B} + \mathbf{C}) = \mathbf{A} \times \mathbf{B} + \mathbf{A} \times \mathbf{C}$$

Chapter 10
Curve Fitting

1. EQUATION CORRESPONDING TO EMPIRICAL DATA

In previous chapters we have found equations of curves which satisfy given geometric conditions. In each case all points of the curve were definitely fixed by the prescribed conditions. We now take up a different and more difficult aspect of the problem of finding equations representing known information. The problem is not primarily of geometric interest, but one in which analytic geometry is fruitful in aiding the scientist. Experimental scientists make observations and measurements of various kinds of natural phenomena. Measurements in an investigation often represent two variable quantities which are related. In many situations the study can be advanced through an equation which expresses the relation, or an approximate relation, between the variables in question. The equation can then be used to compute corresponding values of the variables other than those obtained by measurement. The equation is called an **empirical equation**, and the process employed is called **curve fitting.**

Suppose, for example, that various loads are placed at the midpoint of a beam supported at its ends. If for each load we measure the deflection of the beam at its midpoint, then we obtain a series of corresponding values. One value of each pair is the load and the other the deflection produced by the load. The table lists readings where x stands for the load in pounds and y for the deflection in inches. The pairs of values are plotted as points in Fig. 1. The

x	100	120	140	160	180	200
y	0.45	0.55	0.60	0.70	0.80	0.85

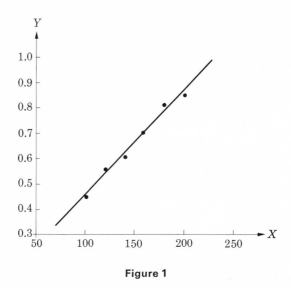

Figure 1

points lie almost in a straight line and suggest that the deflection is proportional to the load; that is, an equation of the form

$$y = mx + b$$

gives the relation, or an approximate relation, between the load and the deflection. Since the points are not exactly in a straight line, no linear equation can be satisfied by all pairs of the readings. We are then faced with the problem of selecting a particular linear equation. A line could be drawn by sight so that it passes quite close to each point. It is desirable, however, to follow some procedure which will locate a definite line. In the next section we shall discuss a method of determining a line which is called the **best fitting line** for a set of data.

If the points representing a set of data are not approximately in a straight line, a linear equation will not express well the relation between the variables; then it would be necessary to seek some nonlinear relation. Although numerous nonlinear equations are used in curve fitting, we shall deal only with the forms

$$y = ax^b$$
$$y = a \cdot 10^{bx}$$
$$y = a \log x + b$$

We shall refer to these equations as the **power formula,** the **exponential formula,** and the **logarithmic formula.** Many physical relations approximately obey one of these types of equations. The equations are advantageous because of their simplicity and because the constants a and b are easily determined.

2. METHOD OF LEAST SQUARES

Suppose that we have given n points in a plane whose coordinates are (x_1, y_1), (x_2, y_2), ..., (x_n, y_n). We define the **residual** of each of the points relative to a curve as the ordinate of the point minus the ordinate of the curve for the same x value. The totality of residuals may be examined to determine if the curve is a good fit to the points. A curve is considered a good fit if each of the residuals is small. Since some of the residuals could be positive and others negative, their sum might be near zero for a curve which is a poor fit to the points. Hence the sum of the residuals would not furnish a reliable measure of the accuracy of fit. For this reason we shall deal with the squares of the residuals, thus avoiding negative quantities. If the sum of the squares of the residuals is small, we would know that the curve passes close to each of the n points. The better fitting of two curves of the same type is the one for which the sum of the squares of the residuals is smaller. The best fitting curve of a given type is the one for which the sum of the squares of the residuals is a minimum.

Starting with the simplest situation, we shall show how to determine the best fitting line to the n given points. We write the linear equation

$$y = mx + b,$$

where values are to be found for m and b so that the sum of the squares of the residuals of the n points is a minimum. The residual of the point (x_1, y_1) is $y_1 - (mx_1 + b)$. The quantity y_1 is the ordinate of the point, and $(mx_1 + b)$ is the ordinate of the line when $x = x_1$. Hence the residuals of the points are

$$y_1 - (mx_1 + b), \; y_2 - (mx_2 + b), \; \ldots, \; y_n - (mx_n + b)$$

and their squares are

$$y_1^2 - 2mx_1 y_1 - 2y_1 b + m^2 x_1^2 + 2mx_1 b + b^2$$
$$y_2^2 - 2mx_2 y_2 - 2y_2 b + m^2 x_2^2 + 2mx_2 b + b^2$$
$$\vdots$$
$$y_n^2 - 2mx_n y_n - 2y_n b + m^2 x_n^2 + 2mx_n b + b^2$$

We use the following convenient notation in finding the sum of these expressions.

$$\sum x \; = x_1 + x_2 + \cdots + x_n$$
$$\sum x^2 = x_1^2 + x_2^2 + \cdots + x_n^2$$
$$\sum xy = x_1 y_1 + x_2 y_2 + \cdots + x_n y_n$$

Denoting the sum of the squares of the residuals by R, we have

$$R = \sum y^2 - 2m \sum xy - 2b \sum y + m^2 \sum x^2 + 2mb \sum x + nb^2 \qquad (1)$$

We note that all quantities in the right member of this equation are fixed in value except for m and b. For example, $\sum y^2$ is not a variable; it stands for the sum of the squares of the ordinates of the n fixed points.

Our problem now is to determine values for m and b which make R a minimum. The expression for R contains the first and second powers of both m and b. If, however, we treat b as an unspecified constant, then the variables in the equation are R and m. Since R appears linearly and m quadratically, the graph of the equation is a parabola. Choosing the m axis as horizontal and the R axis as vertical, we see that the parabola would have a vertical axis. Further, the parabola opens upward because R, being the sum of squared expressions, is not negative. Consequently, the least value of R is the ordinate of the vertex. Hence R takes the least possible value when m is equal to the abscissa of the vertex of the parabola. We can find the abscissa of the vertex by writing Eq. (1) in standard form (Chapter 3, Section 2). In this way it can be shown that R has its least value when

$$m \sum x^2 + b \sum x - \sum xy = 0$$

Similarly, we can consider m as a constant and obtain the equation

$$m \sum x + nb - \sum y = 0$$

Solving the two preceding equations simultaneously for m and b, we obtain

$$m = \frac{n \sum xy - \sum x \sum y}{n \sum x^2 - (\sum x)^2} \qquad b = \frac{\sum x^2 \sum y - \sum x \sum xy}{n \sum x^2 - (\sum x)^2} \qquad (2)$$

These formulas enable us to compute m and b for the line of best fit to a set of given points. We illustrate their use in an example.

Example. Find the line of best fit to the data plotted in Fig. 1.

Solution. The six pairs of corresponding x and y values are listed in Section 1. We use these data to compute the sums appearing in Eq. (2) and obtain

$$\sum x = 100 + 120 + 140 + 160 + 180 + 200 = 900$$
$$\sum y = 0.45 + 0.55 + 0.60 + 0.70 + 0.80 + 0.85 = 3.95$$
$$\sum x^2 = 100^2 + 120^2 + 140^2 + 160^2 + 180^2 + 200^2 = 142{,}000$$
$$\sum xy = 100(.45) + 120(.55) + 140(.60) + 160(.70) + 180(.80) + 200(.85)$$
$$= 621$$

These results, substituted in Eq. (2) for m and b, yield

$$m = \frac{6(621) - 900(3.95)}{6(142,000) - 900^2} = \frac{171}{42,000} = 0.0041$$

$$b = \frac{142,000(3.95) - 900(621)}{42,000} = 0.048$$

Using these values for m and b, we find the equation of the line of best fit to the data to be

$$y = 0.0041x + 0.048$$

This equation gives approximately the relation between the load and deflection and holds for loads which do not bend the beam beyond its elastic limits. The deflection produced by a load of 400 pounds, for example, is $y = 0.0041(400) + 0.048 = 1.69$ inches. The data and the line are shown graphically in Fig. 1.

EXERCISES

Find the equation of the line of best fit to the sets of points in Exercises 1 and 2. Plot the points and draw the line.

1. $(1, 8)$, $(4, 6)$, $(5, 5)$, $(8, 3)$, $(9, 2)$, $(11, 1)$

2. $(-2, -10)$, $(0, -5)$, $(1, 0)$, $(2, 5)$, $(4, 8)$

3. The length y (in.) of a coiled spring under various loads x (lb) are recorded in the table. Find the line of best fit, $y = mx + b$, for these measurements. Use the resulting equation to find the length of the spring when the load is 17 lb.

x	10	20	30	40	50
y	11.0	12.1	13.0	13.9	15.1

4. A business showed net profits at the end of each year for 4 years as follows:

Year	1	2	3	4
Profit	$10,000	$12,000	$13,000	$15,000

Determine the best linear fit and predict the profit for the 5th year.

5. The population N of a city at the end of each decade t for 5 decades is shown in the table. Find the line of best fit, $N = mt + b$, for these data. Predict the population at the end of the 6th decade.

t	1	2	3	4	5
N	8,000	9,000	10,100	11,400	13,700

6. The relation between the total amount of heat H in a pound of saturated steam at T degrees centigrade is $H = mT + b$. Determine m and b for the best linear fit to these data.

T	50	70	90	110
H	623	627	632	636

3. THE POWER FORMULA

Proceeding to nonlinear fits, we consider first the equation

$$y = ax^b$$

The common logarithms of the members of the equation yield

$$\log y = b \log x + \log a$$

This equation, being linear in log x and log y, suggests the plotting of the points (log x, log y). If the points so obtained lie approximately on a line, the power formula is applicable to the set of data. The procedure then is to determine a and b for the best linear fit to the points (log x, log y). The substitution of the values thus determined in the equation $y = ax^b$ gives a best-power fit to the data.

A test of the applicability of a power formula can be made quickly by the use of logarithmic coordinate paper which has its horizontal and vertical rulings placed at distances log 2, log 3, log 4, and so on, from the origin. The original data plotted on this kind of paper are equivalent to plotting the logarithms on regular coordinate paper. The following example illustrates the power-formula method and the use of the special coordinate paper.

Example. The relation between the pressure p and the volume V of a confined gas is given by

$$p = aV^b$$

when the gas neither receives nor loses heat. Determine a and b for the data contained in the table.

V (cu ft)	9.80	6.72	4.53	4.16	3.36	2.83
p (lb/in²)	3	6	9	12	15	18

Solution. The given data are plotted on logarithmic paper in Fig. 2. The points are approximately in a straight line, and therefore indicate that an equation of the power type is suitable for representing the data.

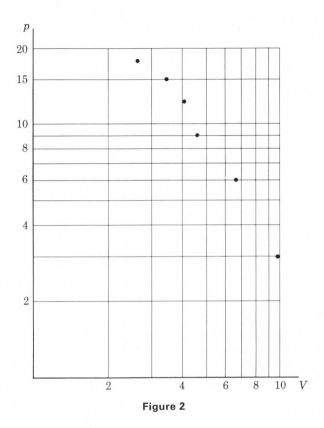

Figure 2

We form the following table by replacing each number of the above data by its common logarithm.

$\log V$	0.9912	0.8274	0.6561	0.6191	0.5263	0.4518
$\log p$	0.4771	0.7782	0.9542	1.0792	1.1761	1.2553

The equation corresponding to this transformed data is

$$\log p = b \log V + \log a$$

To obtain the best linear fit to the new data, we employ Eq. (2), Section 2, using $\log V$ for x and $\log p$ for y. The first formula yields the coefficient of $\log V$

and the second yields the constant term $\log a$. We have the following sums:

$$\sum \log V = 4.0719$$

$$\sum (\log V)^2 = 2.9619$$

$$\sum \log p = 5.7201$$

$$\sum (\log V)(\log p) = 3.5971$$

Substituting these values and $n = 6$, we get

$$b = \frac{6(3.5971) - 4.0719(5.7201)}{6(2.9619) - (4.0719)^2} = -1.43$$

$$\log a = \frac{2.9619(5.7201) - 4.0719(3.5961)}{6(2.9619) - (4.0719)^2} = 1.9272$$

$$a = 84.6$$

Making these substitutions for a and b, we have

$$p = 84.6 \cdot V^{-1.43}$$

The graph of this equation and the points representing the original data are shown in Fig. 3.

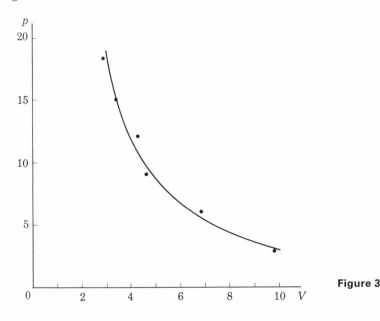

Figure 3

EXERCISES

In Exercises 1 and 2 assume the form $y = ax^b$ and determine a and b for the best fit.

1.

x	2	3	5	7
y	1	4	20	50

2.

x	1	2	3	4
y	0.5	3.0	6.8	10.0

3. A body falls s feet in t seconds. Show that the form $t = as^b$ is applicable to the recorded data and determine a and b for the best fit.

s	4	10	16	25	36
t	0.51	0.79	1.01	1.24	1.49

4. If R is the air resistance in pounds against an automobile traveling at V miles per hour, show that the form $R = aV^b$ is applicable to the measurements in the table and find a and b for the best fit.

V	10	20	30	40	50
R	7	24	65	120	180

5. Corresponding measurements of the volume and pressure of steam are given in the table. Find the best fit of the form $p = aV^n$ to these data.

V	9	5	2.4	2.1	1
p	5	10	30	40	100

4. EXPONENTIAL AND LOGARITHMIC FORMULAS

We saw in the preceding section that the power form can be reduced to a linear form. Similarly, the exponential and logarithmic forms are reducible to linear forms.

We take the common logarithm forms of each member of the equation

$$y = a \cdot 10^{bx}$$

and get

$$\log y = bx + \log a$$

Here $\log y$ is expressed linearly in terms of x. Hence the exponential formula is applicable to a set of data if the points $(x, \log y)$ are in close proximity to a straight line. When this occurs, the procedure is to determine a and b so that $bx + \log a$ is the best linear fit to the set of points $(x, \log y)$.

Semilogarithmic paper may be used to determine whether the exponential formula is adequate to represent the given data. This paper has the usual scale along the x axis and the logarithmic scale along the positive y axis.

Passing finally to the logarithmic formula

$$y = a \log x + b,$$

we see that the equation is linear in y and $\log x$. We consider the points $(\log x, y)$. If these are about in a straight line, then a and b should be found for a linear fit to the points. The values thus obtained should be substituted in the logarithmic equation.

Example. The number of bacteria N per unit volume in a culture after t hours is given by the table for several values of t. Show that $N = a \cdot 10^{bt}$ may represent the data and find values for a and b.

t	1	2	3	4	5
N	70	88	111	127	160

Solution. The given data, plotted on semilogarithmic paper, yield points almost collinear (Fig. 4). This indicates that the data can be approximated by an

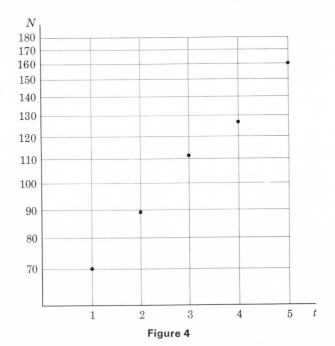

Figure 4

exponential formula. Hence we transform the given data by taking the logarithm of each N.

t	1	2	3	4	5
$\log N$	1.845	1.945	2.045	2.104	2.204

We compute the following sums:

$$\sum t = 15 \qquad\qquad \sum t^2 = 55$$

$$\sum \log N = 10.143 \qquad \sum t(\log N) = 31.306$$

Using these values in Eq. (2), Section 2, we get

$$b = \frac{5(31.306) - 15(10.143)}{5(55) - 15^2} = \frac{4.385}{50} = 0.0877$$

$$\log a = \frac{55(10.143) - 15(31.306)}{50} = 1.766$$

$$a = 58.3$$

We obtain the equation $N = 58.3 \cdot 10^{0.0877t}$. The graph and the given data are shown in Fig. 5.

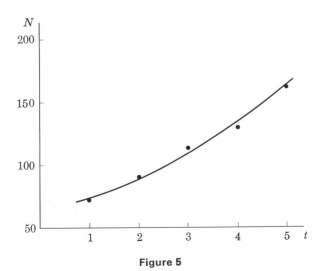

Figure 5

EXERCISES

In Exercises 1 through 3 find best fits of the type indicated.

1.

x	-3	-1	1	3	5
y	0.8	1.5	2.7	4.9	9.0

$y = a \cdot 10^{bx}$

2.

x	0	1	2	3	5
y	3.0	2.5	2.1	1.6	1.1

$y = a \cdot 10^{bx}$

3.

x	$\frac{1}{2}$	1	3	5	8
y	0	3.1	7.8	9.9	12

$y = a \log x + b$

4. The bacteria count N per unit volume in a certain culture at the end of t hours was estimated as in the table. Find the best relation of the form $N = a \cdot 10^{bt}$.

t	0	2	4	6	8
N	10	16	25	40	63

5. The temperature T (degrees C) of a cooling body at time t (min) was measured as recorded. Find an exponential formula of best fit for T in terms of t.

t	0	1	2	3	4	5
T	100	79	63	50	40	32

6. The atmospheric pressure p in pounds per square inch at a height h in thousands of feet is shown in the table. Express p exponentially in terms of h.

h	0	5	10	15	20
p	14.6	12.1	10.1	8.4	7.0

7. The horsepower P required for the speeds V in knots for a certain ship are recorded in the table. Find the best fit to the data of the form $V = a \log P + b$.

P	2000	4000	7000	12000
V	12	13	14	15

Appendix

Table I

POWERS AND ROOTS

No.	Sq.	Sq. Root	Cube	Cube Root	No.	Sq.	Sq. Root	Cube	Cube Root
1	1	1.000	1	1.000	51	2,601	7.141	132,651	3.708
2	4	1.414	8	1.260	52	2,704	7.211	140,608	3.733
3	9	1.732	27	1.442	53	2,809	7.280	148,877	3.756
4	16	2.000	64	1.587	54	2,916	7.348	157,464	3.780
5	25	2.236	125	1.710	55	3,025	7.416	166,375	3.803
6	36	2.449	216	1.817	56	3,136	7.483	175,616	3.826
7	49	2.646	343	1.913	57	3,249	7.550	185,193	3.849
8	64	2.828	512	2.000	58	3,364	7.616	195,112	3.871
9	81	3.000	729	2.080	59	3,481	7.681	205,379	3.893
10	100	3.162	1,000	2.154	60	3,600	7.746	216,000	3.915
11	121	3.317	1,331	2.224	61	3,721	7.810	226,981	3.936
12	144	3.464	1,728	2.289	62	3,844	7.874	238,328	3.958
13	169	3.606	2,197	2.351	63	3,969	7.937	250,047	3.979
14	196	3.742	2,744	2.410	64	4,096	8.000	262,144	4.000
15	225	3.873	3,375	2.466	65	4,225	8.062	274,625	4.021
16	256	4.000	4,096	2.520	66	4,356	8.124	287,496	4.041
17	289	4.123	4,913	2.571	67	4,489	8.185	300,763	4.062
18	324	4.243	5,832	2.621	68	4,624	8.246	314,432	4.082
19	361	4.359	6,859	2.668	69	4,761	8.307	328,509	4.102
20	400	4.472	8,000	2.714	70	4,900	8.367	343,000	4.121
21	441	4.583	9,261	2.759	71	5,041	8.426	357,911	4.141
22	484	4.690	10,648	2.802	72	5,184	8.485	373,248	4.160
23	529	4.796	12,167	2.844	73	5,329	8.544	389,017	4.179
24	576	4.899	13,824	2.884	74	5,476	8.602	405,224	4.198
25	625	5.000	15,625	2.924	75	5,625	8.660	421,875	4.217
26	676	5.099	17,576	2.962	76	5,776	8.718	438,976	4.236
27	729	5.196	19,683	3.000	77	5,929	8.775	456,533	4.254
28	784	5.292	21,952	3.037	78	6,084	8.832	474,552	4.273
29	841	5.385	24,389	3.072	79	6,241	8.888	493,039	4.291
30	900	5.477	27,000	3.107	80	6,400	8.944	512,000	4.309
31	961	5.568	29,791	3.141	81	6,561	9.000	531,441	4.327
32	1,024	5.657	32,768	3.175	82	6,724	9.055	551,368	4.344
33	1,089	5.745	35,937	3.208	83	6,889	9.110	571,787	4.362
34	1,156	5.831	39,304	3.240	84	7,056	9.165	592,704	4.380
35	1,225	5.916	42,875	3.271	85	7,225	9.220	614,125	4.397
36	1,296	6.000	46,656	3.302	86	7,396	9.274	636,056	4.414
37	1,369	6.083	50,653	3.332	87	7,569	9.327	658,503	4.431
38	1,444	6.164	54,872	3.362	88	7,744	9.381	681,472	4.448
39	1,521	6.245	59,319	3.391	89	7,921	9.434	704,969	4.465
40	1,600	6.325	64,000	3.420	90	8,100	9.487	729,000	4.481
41	1,681	6.403	68,921	3.448	91	8,281	9.539	753,571	4.498
42	1,764	6.481	74,088	3.476	92	8,464	9.592	778,688	4.514
43	1,849	6.557	79,507	3.503	93	8,649	9.644	804,357	4.531
44	1,936	6.633	85,184	3.530	94	8,836	9.695	830,584	4.547
45	2,025	6.708	91,125	3.557	95	9,025	9.747	857,375	4.563
46	2,116	6.782	97,336	3.583	96	9,216	9.798	884,736	4.579
47	2,209	6.856	103,823	3.609	97	9,409	9.849	912,673	4.595
48	2,304	6.928	110,592	3.634	98	9,604	9.899	941,192	4.610
49	2,401	7.000	117,649	3.659	99	9,801	9.950	970,299	4.626
50	2,500	7.071	125,000	3.684	100	10,000	10.000	1,000,000	4.642

Table II

NATURAL TRIGONOMETRIC FUNCTIONS

Angle					Angle				
De-gree	Ra-dian	Sine	Co-sine	Tan-gent	De-gree	Ra-dian	Sine	Co-sine	Tan-gent
0°	0.000	0.000	1.000	0.000					
1°	0.017	0.017	1.000	0.017	46°	0.803	0.719	0.695	1.036
2°	0.035	0.035	0.999	0.035	47°	0.820	0.731	0.682	1.072
3°	0.052	0.052	0.999	0.052	48°	0.838	0.743	0.669	1.111
4°	0.070	0.070	0.998	0.070	49°	0.855	0.755	0.656	1.150
5°	0.087	0.087	0.996	0.087	50°	0.873	0.766	0.643	1.192
6°	0.105	0.105	0.995	0.105	51°	0.890	0.777	0.629	1.235
7°	0.122	0.122	0.993	0.123	52°	0.908	0.788	0.616	1.280
8°	0.140	0.139	0.990	0.141	53°	0.925	0.799	0.602	1.327
9°	0.157	0.156	0.988	0.158	54°	0.942	0.809	0.588	1.376
10°	0.175	0.174	0.985	0.176	55°	0.960	0.819	0.574	1.428
11°	0.192	0.191	0.982	0.194	56°	0.977	0.829	0.559	1.483
12°	0.209	0.208	0.978	0.213	57°	0.995	0.839	0.545	1.540
13°	0.227	0.225	0.974	0.231	58°	1.012	0.848	0.530	1.600
14°	0.244	0.242	0.970	0.249	59°	1.030	0.857	0.515	1.664
15°	0.262	0.259	0.966	0.268	60°	1.047	0.866	0.500	1.732
16°	0.279	0.276	0.961	0.287	61°	1.065	0.875	0.485	1.804
17°	0.297	0.292	0.956	0.306	62°	1.082	0.883	0.469	1.881
18°	0.314	0.309	0.951	0.325	63°	1.100	0.891	0.454	1.963
19°	0.332	0.326	0.946	0.344	64°	1.117	0.899	0.438	2.050
20°	0.349	0.342	0.940	0.364	65°	1.134	0.906	0.423	2.145
21°	0.367	0.358	0.934	0.384	66°	1.152	0.914	0.407	2.246
22°	0.384	0.375	0.927	0.404	67°	1.169	0.921	0.391	2.356
23°	0.401	0.391	0.921	0.424	68°	1.187	0.927	0.375	2.475
24°	0.419	0.407	0.914	0.445	69°	1.204	0.934	0.358	2.605
25°	0.436	0.423	0.906	0.466	70°	1.222	0.940	0.342	2.748
26°	0.454	0.438	0.899	0.488	71°	1.239	0.946	0.326	2.904
27°	0.471	0.454	0.891	0.510	72°	1.257	0.951	0.309	3.078
28°	0.489	0.469	0.883	0.532	73°	1.274	0.956	0.292	3.271
29°	0.506	0.485	0.875	0.554	74°	1.292	0.961	0.276	3.487
30°	0.524	0.500	0.866	0.577	75°	1.309	0.966	0.259	3.732
31°	0.541	0.515	0.857	0.601	76°	1.326	0.970	0.242	4.011
32°	0.559	0.530	0.848	0.625	77°	1.344	0.974	0.225	4.332
33°	0.576	0.545	0.839	0.649	78°	1.361	0.978	0.208	4.705
34°	0.593	0.559	0.829	0.675	79°	1.379	0.982	0.191	5.145
35°	0.611	0.574	0.819	0.700	80°	1.396	0.985	0.174	5.671
36°	0.628	0.588	0.809	0.727	81°	1.414	0.988	0.156	6.314
37°	0.646	0.602	0.799	0.754	82°	1.431	0.990	0.139	7.115
38°	0.663	0.616	0.788	0.781	83°	1.449	0.993	0.122	8.144
39°	0.681	0.629	0.777	0.810	84°	1.466	0.995	0.105	9.514
40°	0.698	0.643	0.766	0.839	85°	1.484	0.996	0.087	11.43
41°	0.716	0.656	0.755	0.869	86°	1.501	0.998	0.070	14.30
42°	0.733	0.669	0.743	0.900	87°	1.518	0.999	0.052	19.08
43°	0.750	0.682	0.731	0.933	88°	1.536	0.999	0.035	28.64
44°	0.768	0.695	0.719	0.966	89°	1.553	1.000	0.017	57.29
45°	0.785	0.707	0.707	1.000	90°	1.571	1.000	0.000	

Table III

EXPONENTIAL FUNCTIONS

x	e^x	e^{-x}	x	e^x	e^{-x}
0.00	1.0000	1.0000	2.5	12.182	0.0821
0.05	1.0513	0.9512	2.6	13.464	0.0743
0.10	1.1052	0.9048	2.7	14.880	0.0672
0.15	1.1618	0.8607	2.8	16.445	0.0608
0.20	1.2214	0.8187	2.9	18.174	0.0550
0.25	1.2840	0.7788	3.0	20.086	0.0498
0.30	1.3499	0.7408	3.1	22.198	0.0450
0.35	1.4191	0.7047	3.2	24.533	0.0408
0.40	1.4918	0.6703	3.3	27.113	0.0369
0.45	1.5683	0.6376	3.4	29.964	0.0334
0.50	1.6487	0.6065	3.5	33.115	0.0302
0.55	1.7333	0.5769	3.6	36.598	0.0273
0.60	1.8221	0.5488	3.7	40.447	0.0247
0.65	1.9155	0.5220	3.8	44.701	0.0224
0.70	2.0138	0.4966	3.9	49.402	0.0202
0.75	2.1170	0.4724	4.0	54.598	0.0183
0.80	2.2255	0.4493	4.1	60.340	0.0166
0.85	2.3396	0.4274	4.2	66.686	0.0150
0.90	2.4596	0.4066	4.3	73.700	0.0136
0.95	2.5857	0.3867	4.4	81.451	0.0123
1.0	2.7183	0.3679	4.5	90.017	0.0111
1.1	3.0042	0.3329	4.6	99.484	0.0101
1.2	3.3201	0.3012	4.7	109.95	0.0091
1.3	3.6693	0.2725	4.8	121.51	0.0082
1.4	4.0552	0.2466	4.9	134.29	0.0074
1.5	4.4817	0.2231	5	148.41	0.0067
1.6	4.9530	0.2019	6	403.43	0.0025
1.7	5.4739	0.1827	7	1096.6	0.0009
1.8	6.0496	0.1653	8	2981.0	0.0003
1.9	6.6859	0.1496	9	8103.1	0.0001
2.0	7.3891	0.1353	10	22026	0.00005
2.1	8.1662	0.1225			
2.2	9.0250	0.1108			
2.3	9.9742	0.1003			
2.4	11.023	0.0907			

Table IV

COMMON LOGARITHMS OF NUMBERS

N	0	1	2	3	4	5	6	7	8	9
0	....	0000	3010	4771	6021	6990	7782	8451	9031	9542
1	0000	0414	0792	1139	1461	1761	2041	2304	2553	2788
2	3010	3222	3424	3617	3802	3979	4150	4314	4472	4624
3	4771	4914	5051	5185	5315	5441	5563	5682	5798	5911
4	6021	6128	6232	6335	6435	6532	6628	6721	6812	6902
5	6990	7076	7160	7243	7324	7404	7482	7559	7634	7709
6	7782	7853	7924	7993	8062	8129	8195	8261	8325	8388
7	8451	8513	8573	8633	8692	8751	8808	8865	8921	8976
8	9031	9085	9138	9191	9243	9294	9345	9395	9445	9494
9	9542	9590	9638	9685	9731	9777	9823	9868	9912	9956
10	0000	0043	0086	0128	0170	0212	0253	0294	0334	0374
11	0414	0453	0492	0531	0569	0607	0645	0682	0719	0755
12	0792	0828	0864	0899	0934	0969	1004	1038	1072	1106
13	1139	1173	1206	1239	1271	1303	1335	1367	1399	1430
14	1461	1492	1523	1553	1584	1614	1644	1673	1703	1732
15	1761	1790	1818	1847	1875	1903	1931	1959	1987	2014
16	2041	2068	2095	2122	2148	2175	2201	2227	2253	2279
17	2304	2330	2355	2380	2405	2430	2455	2480	2504	2529
18	2553	2577	2601	2625	2648	2672	2695	2718	2742	2765
19	2788	2810	2833	2856	2878	2900	2923	2945	2967	2989
20	3010	3032	3054	3075	3096	3118	3139	3160	3181	3201
21	3222	3243	3263	3284	3304	3324	3345	3365	3385	3404
22	3424	3444	3464	3483	3502	3522	3541	3560	3579	3598
23	3617	3636	3655	3674	3692	3711	3729	3747	3766	3784
24	3802	3820	3838	3856	3874	3892	3909	3927	3945	3962
25	3979	3997	4014	4031	4048	4065	4082	4099	4116	4133
26	4150	4166	4183	4200	4216	4232	4249	4265	4281	4298
27	4314	4330	4346	4362	4378	4393	4409	4425	4440	4456
28	4472	4487	4502	4518	4533	4548	4564	4579	4594	4609
29	4624	4639	4654	4669	4683	4698	4713	4728	4742	4757
30	4771	4786	4800	4814	4829	4843	4857	4871	4886	4900
31	4914	4928	4942	4955	4969	4983	4997	5011	5024	5038
32	5051	5065	5079	5092	5105	5119	5132	5145	5159	5172
33	5185	5198	5211	5224	5237	5250	5263	5276	5289	5302
34	5315	5328	5340	5353	5366	5378	5391	5403	5416	5428
35	5441	5453	5465	5478	5490	5502	5514	5527	5539	5551
36	5563	5575	5587	5599	5611	5623	5635	5647	5658	5670
37	5682	5694	5705	5717	5729	5740	5752	5763	5775	5786
38	5798	5809	5821	5832	5843	5855	5866	5877	5888	5899
39	5911	5922	5933	5944	5955	5966	5977	5988	5999	6010
40	6021	6031	6042	6053	6064	6075	6085	6096	6107	6117
41	6128	6138	6149	6160	6170	6180	6191	6201	6212	6222
42	6232	6243	6253	6263	6274	6284	6294	6304	6314	6325
43	6335	6345	6355	6365	6375	6385	6395	6405	6415	6425
44	6435	6444	6454	6464	6474	6484	6493	6503	6513	6522
45	6532	6542	6551	6561	6571	6580	6590	6599	6609	6618
46	6628	6637	6646	6656	6665	6675	6684	6693	6702	6712
47	6721	6730	6739	6749	6758	6767	6776	6785	6794	6803
48	6812	6821	6830	6839	6848	6857	6866	6875	6884	6893
49	6902	6911	6920	6928	6937	6946	6955	6964	6972	6981
50	6990	6998	7007	7016	7024	7033	7042	7050	7059	7067
N	0	1	2	3	4	5	6	7	8	9

Table IV

Common Logarithms of Numbers

N	0	1	2	3	4	5	6	7	8	9
50	6990	6998	7007	7016	7024	7033	7042	7050	7059	7067
51	7076	7084	7093	7101	7110	7118	7126	7135	7143	7152
52	7160	7168	7177	7185	7193	7202	7210	7218	7226	7235
53	7243	7251	7259	7267	7275	7284	7292	7300	7308	7316
54	7324	7332	7340	7348	7356	7364	7372	7380	7388	7396
55	7404	7412	7419	7427	7435	7443	7451	7459	7466	7474
56	7482	7490	7497	7505	7513	7520	7528	7536	7543	7551
57	7559	7566	7574	7582	7589	7597	7604	7612	7619	7627
58	7634	7642	7649	7657	7664	7672	7679	7686	7694	7701
59	7709	7716	7723	7731	7738	7745	7752	7760	7767	7774
60	7782	7789	7796	7803	7810	7818	7825	7832	7839	7846
61	7853	7860	7868	7875	7882	7889	7896	7903	7910	7917
62	7924	7931	7938	7945	7952	7959	7966	7973	7980	7987
63	7993	8000	8007	8014	8021	8028	8035	8041	8048	8055
64	8062	8069	8075	8082	8089	8096	8102	8109	8116	8122
65	8129	8136	8142	8149	8156	8162	8169	8176	8182	8189
66	8195	8202	8209	8215	8222	8228	8235	8241	8248	8254
67	8261	8267	8274	8280	8287	8293	8299	8306	8312	8319
68	8325	8331	8338	8344	8351	8357	8363	8370	8376	8382
69	8388	8395	8401	8407	8414	8420	8426	8432	8439	8445
70	8451	8457	8463	8470	8476	8482	8488	8494	8500	8506
71	8513	8519	8525	8531	8537	8543	8549	8555	8561	8567
72	8573	8579	8585	8591	8597	8603	8609	8615	8621	8627
73	8633	8639	8645	8651	8657	8663	8669	8675	8681	8686
74	8692	8698	8704	8710	8716	8722	8727	8733	8739	8745
75	8751	8756	8762	8768	8774	8779	8785	8791	8797	8802
76	8808	8814	8820	8825	8831	8837	8842	8848	8854	8859
77	8865	8871	8876	8882	8887	8893	8899	8904	8910	8915
78	8921	8927	8932	8938	8943	8949	8954	8960	8965	8971
79	8976	8982	8987	8993	8998	9004	9009	9015	9020	9025
80	9031	9036	9042	9047	9053	9058	9063	9069	9074	9079
81	9085	9090	9096	9101	9106	9112	9117	9122	9128	9133
82	9138	9143	9149	9154	9159	9165	9170	9175	9180	9186
83	9191	9196	9201	9206	9212	9217	9222	9227	9232	9238
84	9243	9248	9253	9258	9263	9269	9274	9279	9284	9289
85	9294	9299	9304	9309	9315	9320	9325	9330	9335	9340
86	9345	9350	9355	9360	9365	9370	9375	9380	9385	9390
87	9395	9400	9405	9410	9415	9420	9425	9430	9435	9440
88	9445	9450	9455	9460	9465	9469	9474	9479	9484	9489
89	9494	9499	9504	9509	9513	9518	9523	9528	9533	9538
90	9542	9547	9552	9557	9562	9566	9571	9576	9581	9586
91	9590	9595	9600	9605	9609	9614	9619	9624	9628	9633
92	9638	9643	9647	9652	9657	9661	9666	9671	9675	9680
93	9685	9689	9694	9699	9703	9708	9713	9717	9722	9727
94	9731	9736	9741	9745	9750	9754	9759	9763	9768	9773
95	9777	9782	9786	9791	9795	9800	9805	9809	9814	9818
96	9823	9827	9832	9836	9841	9845	9850	9854	9859	9863
97	9868	9872	9877	9881	9886	9890	9894	9899	9903	9908
98	9912	9917	9921	9926	9930	9934	9939	9943	9948	9952
99	9956	9961	9965	9969	9974	9978	9983	9987	9991	9996
100	0000	0004	0009	0013	0017	0022	0026	0030	0035	0039
N	0	1	2	3	4	5	6	7	8	9

Table V

Natural Logarithms of Numbers

n	$\log_e n$	n	$\log_e n$	n	$\log_e n$
0.0	*	4.5	1.5041	9.0	2.1972
0.1	7.6974	4.6	1.5261	9.1	2.2083
0.2	8.3906	4.7	1.5476	9.2	2.2192
0.3	8.7960	4.8	1.5686	9.3	2.2300
0.4	9.0837	4.9	1.5892	9.4	2.2407
0.5	9.3069	5.0	1.6094	9.5	2.2513
0.6	9.4892	5.1	1.6292	9.6	2.2618
0.7	9.6433	5.2	1.6487	9.7	2.2721
0.8	9.7769	5.3	1.6677	9.8	2.2824
0.9	9.8946	5.4	1.6864	9.9	2.2925
1.0	0.0000	5.5	1.7047	10	2.3026
1.1	0.0953	5.6	1.7228	11	2.3979
1.2	0.1823	5.7	1.7405	12	2.4849
1.3	0.2624	5.8	1.7579	13	2.5649
1.4	0.3365	5.9	1.7750	14	2.6391
1.5	0.4055	6.0	1.7918	15	2.7081
1.6	0.4700	6.1	1.8083	16	2.7726
1.7	0.5306	6.2	1.8245	17	2.8332
1.8	0.5878	6.3	1.8405	18	2.8904
1.9	0.6419	6.4	1.8563	19	2.9444
2.0	0.6931	6.5	1.8718	20	2.9957
2.1	0.7419	6.6	1.8871	25	3.2189
2.2	0.7885	6.7	1.9021	30	3.4012
2.3	0.8329	6.8	1.9169	35	3.5553
2.4	0.8755	6.9	1.9315	40	3.6889
2.5	0.9163	7.0	1.9459	45	3.8067
2.6	0.9555	7.1	1.9601	50	3.9120
2.7	0.9933	7.2	1.9741	55	4.0073
2.8	1.0296	7.3	1.9879	60	4.0943
2.9	1.0647	7.4	2.0015	65	4.1744
3.0	1.0986	7.5	2.0149	70	4.2485
3.1	1.1314	7.6	2.0281	75	4.3175
3.2	1.1632	7.7	2.0412	80	4.3820
3.3	1.1939	7.8	2.0541	85	4.4427
3.4	1.2238	7.9	2.0669	90	4.4998
3.5	1.2528	8.0	2.0794	95	4.5539
3.6	1.2809	8.1	2.0919	100	4.6052
3.7	1.3083	8.2	2.1041		
3.8	1.3350	8.3	2.1163		
3.9	1.3610	8.4	2.1282		
4.0	1.3863	8.5	2.1401		
4.1	1.4110	8.6	2.1518		
4.2	1.4351	8.7	2.1633		
4.3	1.4586	8.8	2.1748		
4.4	1.4816	8.9	2.1861		

Formulas

ALGEBRA

1. Quadratic formula. The roots of the equation

$$ax^2 + bx + c = 0 \qquad (a \neq 0)$$

are

$$x = \frac{-b \pm \sqrt{b^2 - 4ac}}{2a}.$$

2. Logarithms. If a is a positive number different from 1 and y is any real number, then

$$y = \log_a x$$

if and only if

$$a^y = x.$$

3. Change of base for logarithms

$$\log_a x = \frac{\log_b x}{\log_b a}; \qquad \log_{10} x = \frac{\log_e x}{\log_e 10}; \qquad \log_e x = \frac{\log_{10} x}{\log_{10} e}.$$

TRIGONOMETRY

4. Definitions and fundamental identities

$$\sin \theta = \frac{y}{r} = \frac{1}{\csc \theta};$$

$$\cos \theta = \frac{x}{r} = \frac{1}{\sec \theta};$$

$$\tan \theta = \frac{y}{x} = \frac{1}{\cot \theta} = \frac{\sin \theta}{\cos \theta};$$

$$\sin^2 \theta + \cos^2 \theta = 1;$$
$$1 + \tan^2 \theta = \sec^2 \theta;$$
$$1 + \cot^2 \theta = \csc^2 \theta;$$

$$\sin \theta = \cos (90 - \theta) = \sin (180 - \theta);$$
$$\cos \theta = \sin (90 - \theta) = -\cos (180 - \theta);$$
$$\tan \theta = \cot (90 - \theta) = -\tan (180 - \theta);$$
$$\cot \theta = \tan (90 - \theta) = -\cot (180 - \theta);$$

$$\csc \theta = \cot \frac{\theta}{2} - \cot \theta;$$

$\sin (\theta + \phi) = \sin \theta \cos \phi + \cos \theta \sin \phi;$
$\sin (\theta - \phi) = \sin \theta \cos \phi - \cos \theta \sin \phi;$
$\cos (\theta + \phi) = \cos \theta \cos \phi - \sin \theta \sin \phi;$
$\cos (\theta - \phi) = \cos \theta \cos \phi + \sin \theta \sin \phi;$

$$\tan (\theta + \phi) = \frac{\tan \theta + \tan \phi}{1 - \tan \theta \tan \phi};$$

$$\tan (\theta - \phi) = \frac{\tan \theta - \tan \phi}{1 + \tan \theta \tan \phi};$$

$\sin 2\theta = 2 \sin \theta \cos \theta;$
$\cos 2\theta = \cos^2 \theta - \sin^2 \theta = 2 \cos^2 \theta - 1 = 1 - 2 \sin^2 \theta;$

$$\tan 2\theta = \frac{2 \tan \theta}{1 - \tan^2 \theta}; \qquad \cot 2\theta = \frac{\cot^2 \theta - 1}{2 \cot \theta};$$

$$\sin \frac{\theta}{2} = \pm \sqrt{\frac{1 - \cos \theta}{2}};$$

$$\cos \frac{\theta}{2} = \pm \sqrt{\frac{1 + \cos \theta}{2}};$$

$$\tan \frac{\theta}{2} = \pm \sqrt{\frac{1 - \cos \theta}{1 + \cos \theta}} = \frac{1 - \cos \theta}{\sin \theta} = \frac{\sin \theta}{1 + \cos \theta};$$

$$\sin \theta + \sin \phi = 2 \sin \frac{\theta + \phi}{2} \cos \frac{\theta - \phi}{2};$$

$$\sin \theta - \sin \phi = 2 \cos \frac{\theta + \phi}{2} \sin \frac{\theta - \phi}{2};$$

$$\cos \theta + \cos \phi = 2 \cos \frac{\theta + \phi}{2} \cos \frac{\theta - \phi}{2};$$

$$\cos \theta - \cos \phi = -2 \sin \frac{\theta + \phi}{2} \sin \frac{\theta - \phi}{2};$$

$\sin \theta \cos \phi = \frac{1}{2}[\sin (\theta + \phi) + \sin (\theta - \phi)];$
$\cos \theta \cos \phi = \frac{1}{2}[\cos (\theta + \phi) + \cos (\theta - \phi)];$
$\sin \theta \sin \phi = -\frac{1}{2}[\cos (\theta + \phi) - \cos (\theta - \phi)].$

5. Angles and sides of triangles

Law of cosines: $a^2 = b^2 + c^2 - 2bc \cos A.$

Law of sines: $\dfrac{\sin A}{a} = \dfrac{\sin B}{b} = \dfrac{\sin C}{c}.$

Area $= \frac{1}{2}bc \sin A = \frac{1}{2}ac \sin B = \frac{1}{2}ab \sin C.$

ANALYTIC GEOMETRY

6. Basic formulas

Distance between two points: $P_1P_2 = \sqrt{(x_2 - x_1)^2 + (y_2 - y_1)^2}$.

Midpoint: $P\left(\dfrac{x_1 + x_2}{2}, \dfrac{y_1 + y_2}{2}\right)$.

Slope of a line: $m = \dfrac{y_2 - y_1}{x_2 - x_1} = \dfrac{y_1 - y_2}{x_1 - x_2}$.

Condition for parallel lines: $m_1 = m_2$.
Condition for perpendicular lines: $m_1 m_2 = -1$.

Angle between two lines: $\tan \theta = \dfrac{m_2 - m_1}{1 + m_2 m_1}$.

Distance between a point and a line: $PQ = \dfrac{|Ax_1 + By_1 + C|}{\sqrt{A^2 + B^2}}$.

7. Equation of straight line

General equation: $Ax + By + C = 0$.

Two point form: $y - y_1 = \dfrac{y_2 - y_1}{x_2 - x_1}(x - x_1)$.

Point-slope form: $y - y_1 = m(x - x_1)$.
Slope-intercept form: $y = mx + b$.

Intercept form: $\dfrac{x}{a} + \dfrac{y}{b} = 1$.

8. Circle

General equation: $x^2 + y^2 + Dx + Ey + F = 0$.
Center at origin, radius r: $x^2 + y^2 = r^2$, or $x = r \cos \theta$,
$y = r \sin \theta$.
Equation of line tangent at (x_1, y_1): $x_1 x + y_1 y = r^2$.
Center at (h, k), radius r: $(x - h)^2 + (y - k)^2 = r^2$.

9. Parabola

Vertex at origin, opening in direction of positive x:

$$y^2 = 4ax(a > 0), \text{ or } x = at^2, y = 2at.$$

Equation of line tangent at (x_1, y_1): $y_1 y = 2a(x - x_1)$.
Focus: $(a, 0)$.   Directrix: $x + a = 0$.
Vertex at (h, k), opening in direction of positive x:

$$(y - k)^2 = 4a(x - h).$$

10. Ellipse

Center at origin: $\dfrac{x^2}{a^2} + \dfrac{y^2}{b^2} = 1$, or $x = a\cos\theta$, $y = b\sin\theta$.

Equation of line tangent at (x_1, y_1): $\dfrac{x_1 x}{a^2} + \dfrac{y_1 y}{b^2} = 1$.

If $a > b$: $c^2 = a^2 - b^2$; eccentricity $e = c/a$; foci, $(-c, \theta)$, (c, θ).
If $b > a$: $c^2 = b^2 - a^2$; eccentricity $e = c/b$; foci, $(0, -c)$, $(0, c)$.

Center at (h, k): $\dfrac{(x - h)^2}{a^2} + \dfrac{(y - k)^2}{b^2} = 1$.

11. Hyperbola

Center at origin: $\dfrac{x^2}{a^2} - \dfrac{y^2}{b^2} = 1$, or $x = a\sec\theta$, $y = b\tan\theta$.

Equation of line tangent at (x_1, y_1): $\dfrac{x_1 x}{a^2} - \dfrac{y_1 y}{b^2} = 1$.

Asymptotes: $y = \dfrac{b}{a}x$, $y = -\dfrac{b}{a}x$.

Foci: $(-c, 0)$, $(c, 0)$.

Eccentricity: $e = \dfrac{c}{a}$, where $c^2 = a^2 + b^2$.

Center at (h, k): $(x - h)^2/a^2 - (y - k)^2/b^2 = \pm 1$.
With asymptotes $x = a$, $y = b$: $(x - a)(y - b) = k$.

Answers to Odd-Numbered
Exercises

Answers to Odd-Numbered
Exercises

Chapter 1

Section 4, page 9

1. $\overrightarrow{AB} = 3$, $\overrightarrow{AC} = 7$, $\overrightarrow{BC} = 4$, $\overrightarrow{BA} = -3$, $\overrightarrow{CA} = -7$, $\overrightarrow{CB} = -4$

5. 13

7. 13

9. $3\sqrt{2}$

11. $\sqrt{13}$, $\sqrt{10}$, $\sqrt{5}$

13. $3\sqrt{2}$, 1, 5

25. Each point is 5 units from $(-2, 3)$.

27. On a line.

29. Not on a line.

31. $x = 2$

35. 9

37. 23.5

39. 19

Section 6, page 18

5. $1, 0, \sqrt{3}, -\sqrt{3}, -1$

7. $-\frac{8}{3}$, $111°$

9. $\frac{4}{3}$, $53°$

11. $\frac{7}{6}$, $49°$

15. On a line.

17. Not on a line.

19. $\tan A = \frac{11}{16}$, $A = 35°$; $\tan B = \frac{11}{10}$, $B = 48°$; $\tan C = -\frac{22}{3}$, $C = 98°$

21. $\tan A = \frac{9}{37}$, $A = 14°$; $\tan B = \frac{9}{13}$, $B = 35°$; $\tan C = -\frac{9}{8}$, $C = 132°$

23. $85°, 95°$

25. $56°$

27. $\frac{22}{9}$

29. $9x - 8y + 14 = 0$

31. $6x - 5y = 37$

Section 8, page 24

1. a) $(0,0)$; b) $(-2,-1)$; c) $(4,1)$; d) $(7.5, 2.5)$; e) $(-5, 2)$; f) $(-5.5, -6)$

3. $x_1 = 6, y_1 = -3$ 5. $(\frac{5}{2}, \frac{7}{2})$ 7. $(\frac{3}{5}, \frac{16}{5})$

9. $(\frac{13}{4}, -\frac{31}{8})$ 11. $(\frac{11}{2}, 2), (-5, -3)$

Section 11, page 34

1. $y = 2x$ 3. $x - 2y + 8 = 0$ 5. $x = 6$

7. $y = -2$ 9. $x - 3y + 1 = 0$ 11. $y^2 - 10x + 25 = 0$

13. $x^2 + y^2 = 8$ 15. $x^2 + y^2 = 25$ 17. $16x^2 + 25y^2 = 400$

Chapter 2

Section 1, page 39

1. $m = 4, a = 3, b = -12; y = 4x - 12$

3. $m = -1, a = -4, b = -4; y = -x - 4$

5. $m = \frac{3}{4}, a = 4, b = -3; y = \frac{3}{4}x - 3$

7. $m = -\frac{1}{7}, a = 11, b = \frac{11}{7}; y = -\frac{1}{7}x + \frac{11}{7}$

9. $m = -\frac{7}{3}, a = -\frac{6}{7}, b = -2; y = -\frac{7}{3}x - 2$

11. $m = -\frac{8}{3}, a = \frac{1}{2}, b = \frac{4}{3}; y = -\frac{8}{3}x + \frac{4}{3}$

13. $y = 3x - 4$ 15. $y = -4x + 5$ 17. $y = \frac{2}{3}x - 2$

19. $y = 7$ 21. $2x - y = 5$ 23. $2x - 3y + 4 = 0$

25. $x + 2y + 15 = 0$ 27. $y = 3$ 29. $8x + 3y = 0$

31. $6x + 7y = 11$ 33. $2x + y - 2 = 0$ 35. $x = 3$

37. $8x - 27y = 58$ 39. $y + 1 = 0$

Section 2, page 42

1. $2x - 3y + 10 = 0$ 3. $5x + 3y + 20 = 0$

5. $7x + 9y + 51 = 0$ 7. $3x + 2y = 6$

9. $4x - 3y + 12 = 0$ 11. $x + y + 4 = 0$

13. $5x + 8y = 4$ 15. $\dfrac{x}{8} + \dfrac{y}{-2} = 1$

17. $\dfrac{x}{-9} + \dfrac{y}{4} = 1$

19. $\dfrac{x}{9/2} + \dfrac{y}{-9/5} = 1$

21. $\dfrac{x}{5/4} + \dfrac{y}{-5/3} = 1$

23. $\dfrac{x}{2} + \dfrac{y}{3/5} = 1$

25. $2x - 3y = 5,\ 3x + 2y = 14$

27. $8x - y - 19 = 0,\ x + 8y + 22 = 0$

29. $9x + y - 63 = 0,\ x - 9y - 7 = 0$ 31. $y = 1,\ x = -1$

33. a) $2x - 5y = 0,\ x + 2y = 6,\ 4x - y = 12;\ (\frac{10}{3}, \frac{4}{3})$
   b) $x - 2y = 0,\ x + y = 6,\ x = 4;\ (4, 2)$
   c) $x = 3,\ x - 2y = 1,\ x + y = 4;\ (3, 1)$

35. a) $4x - 5y = 4,\ x + 7y = 23,\ 5x + 2y = 27;\ (\frac{13}{3}, \frac{8}{3})$
   b) $3x - 2y = 3,\ x + 3y = 15,\ 4x + y = 18;\ (\frac{39}{11}, \frac{42}{11})$
   c) $4x + y = 21,\ x + 3y = 11,\ 3x - 2y = 10;\ (\frac{52}{11}, \frac{23}{11})$

Section 3, page 48

1. $\frac{19}{3}$ 3. 10.4 5. $-\sqrt{2}$ 7. 3

9. 3 11. 5 13. 8.5 15. $8x - 25 = 0$

17. $99x + 27y - 256 = 0,\ 21x - 77y + 56 = 0$ 19. Area $= 17$

Section 4, page 52

1. $7x - 2y = k$

3. $x + 2y = 2a,\ a \neq 0$

5. $4x + ay = 4a,\ a \neq 0$

7. $32x + a^2 y = 32a$

9. All have slope $= 2$.

11. All pass through $(3, -4)$.

13. All have y intercept $= -2,\ b \neq 0$.

15. All have x intercept $= 4,\ a \neq 0$.

17. $y = 2x - 7,\ x + y + 1 = 0,\ x + 2y + 4 = 0,\ 3x - 2y = 12$

19. $4x + 3y + C = 0,\ 4x + 3y + 29 = 0,\ 4x + 3y - 31 = 0$

21. $3x - y + 6 + 2\sqrt{10} = 0,\ 3x - y + 6 - 2\sqrt{10} = 0$

23. $5x - y + 6 = 0$ 25. $x - 3y = 0$

27. $x + y = 5$ or $3x - 2y = 0$ 29. $34x - 55 = 0$

31. $7x + 7y - 1 = 0,\ x - 2y - 3 = 0,\ 15x - 9y - 25 = 0$

Section 6, page 59

1. $(x - 2)^2 + (y + 6)^2 = 25$

3. $x^2 + (y - 4)^2 = 16$

5. $(x + 12)^2 + (y - 5)^2 = 169$

7. $(x - \frac{1}{2})^2 + (y + 3)^2 = 11$

9. $x^2 + y^2 + 8x - 6y = 0$

11. $x^2 + y^2 - 8x - 4y = 14$

13. $x^2 + y^2 + 8x - 2y + 16 = 0$

15. $x^2 + y^2 - 14y = 95$

17. $(x - 3)^2 + (y + 2)^2 = 25$

19. $(x + 4)^2 + (y + 1)^2 = 16$

21. $(x - 4)^2 + (y - 3)^2 = 25$

23. $(x - 2)^2 + (y + 6)^2 = 48$

25. $(x - 3)^2 + (y + \frac{1}{2})^2 = \frac{35}{4}$

27. Circle

29. The point $(-1, 0)$

31. Circle

33. The point $(-1, -5)$

35. No graph

37. $x^2 + y^2 - 4y - 1 = 0$

39. $(x - 7)^2 + (y - 6)^2 = 26$

41. $(x - 1)^2 + (y + 3)^2 = 25$

43. $x^2 + y^2 - 5x - y = 0$

45. $x^2 + y^2 - 2x + 2y - 23 = 0$

47. $(x - \frac{5}{6})^2 + (y + \frac{4}{3})^2 = \frac{529}{360}$

Section 7, page 64

5. $5x^2 + 5y^2 - 26x - 4y - 25 = 0$

7. $(x - 2)^2 + (y - \frac{9}{4})^2 = \frac{25}{16}$

Section 8, page 67

1. $(1, -1), (-3, -1), (-3, -7), (1, -7)$

3. $(-4, 7), (-2, 8), (-3, 0), (3, 6)$

5. $(1, 8), (-5, 4), (2, 9), (-4, -2)$

7. $x' + 2y' = 0$

9. $x'^2 + y'^2 = 1$

11. $x'^2 + y'^2 = 5$

13. $x'^2 + y'^2 = 64$

15. $(-2, 1), x'^2 + y'^2 = 10$

17. $(2, -3), 4x'^2 + y'^2 = 28$

Chapter 3

Section 1, page 76

1. $(2, 0), 8, (2, 4), (2, -4), x = -2$

3. $(-3, 0), 12, (-3, 6), (-3, -6), x = 3$

5. $(0, -\frac{5}{2}), 10, (-5, -\frac{5}{2}), (5, -\frac{5}{2}), y = \frac{5}{2}$

7. $(\frac{9}{8}, 0), \frac{9}{2}, (\frac{9}{8}, \frac{9}{4}), (\frac{9}{8}, -\frac{9}{4}), x = -\frac{9}{8}$

9. $(0, \frac{9}{4}), 9, (-\frac{9}{2}, \frac{9}{4}), (\frac{9}{2}, \frac{9}{4}), y = -\frac{9}{4}$

11. $x^2 = 8y$

13. $x^2 = -8y$

15. $x^2 = 16y$

17. $4y^2 = 9x$

19. $2y^2 = -x$

21. $x^2 = 200y$

Section 3, page 81

1. $(y - 3)^2 = 16x$

3. $(y - 3)^2 = 16(x - 2)$

5. $(y - 3)^2 = -24(x - 3)$

7. $(x + 1)^2 = -12(y + 2)$

9. $(y - 1)^2 = -12(x - 2)$

11. $y^2 = 8(x-1)$; $V(1,0)$, $F(3,0)$; $(3,-4)$, $(3,4)$

13. $y^2 = -12(x-4)$; $V(4,0)$, $F(1,0)$; $(1,-6)$, $(1,6)$

15. $(x+2)^2 = 16y$; $V(-2,0)$, $F(-2,4)$; $(-10,4)$, $(6,4)$

17. $(y-4)^2 = -6x$; $V(0,4)$, $F(-\frac{3}{2},4)$; $(-\frac{3}{2},1)$, $(-\frac{3}{2},7)$

19. $(y+2)^2 = -8(x-4)$; $V(4,-2)$, $F(2,-2)$; $(2,-6)$, $(2,2)$

21. $(x-4)^2 = -6(y-4)$; $V(4,4)$, $F(4,\frac{5}{2})$; $(1,\frac{5}{2})$, $(7,\frac{5}{2})$

23. $(y+7)^2 = 24(x+7)$; $V(-7,-7)$, $F(-1,-7)$; $(-1,-19)$, $(-1,5)$

25. $(x+1)^2 = 2(y+2)$ 27. $y^2 - x + 2y - 2 = 0$

29. $y^2 + x - 3y - 4 = 0$ 31. $5x^2 + 3x - 6y - 20 = 0$

Section 7, page 91

1. $F(\pm 4, 0)$; $V(\pm 5, 0)$; $B(0, \pm 3)$; $(4, \pm\frac{9}{5})$, $(-4, \pm\frac{9}{5})$

3. $F(0, \pm 3)$; $V(0, \pm 5)$; $B(\pm 4, 0)$; $(\frac{16}{5}, \pm 3)$. $(-\frac{16}{5}, \pm 3)$

5. $F(\pm 2\sqrt{6}, 0)$; $V(\pm 7, 0)$; $B(0, \pm 5)$; $(2\sqrt{6}, \pm\frac{25}{7})$, $(-2\sqrt{6}, \pm\frac{25}{7})$

7. $F(\pm\sqrt{15}, 0)$; $V(\pm 4, 0)$; $B(0, \pm 1)$; $(\sqrt{15}, \pm\frac{1}{4})$, $(-\sqrt{15}, \pm\frac{1}{4})$

9. $F(2 \pm \sqrt{7}, 3)$; $V(2 \pm 4, 3)$; $B(2, 3 \pm 3)$; $(2 + \sqrt{7}, 3 \pm \frac{9}{4})$, $(2 - \sqrt{7}, 3 \pm \frac{9}{4})$

11. $F(5, -5 \pm 2\sqrt{30})$; $V(5, -5 \pm 13)$; $B(5 \pm 7, -5)$; $(5 \pm \frac{49}{13}, -5 + 2\sqrt{30})$,
$(5 \pm \frac{49}{13}, -5 - 2\sqrt{30})$

13. $\dfrac{(x-5)^2}{25} + \dfrac{(y-4)^2}{16} = 1$; center $(5,4)$; $F'(2,4)$, $F(8,4)$; $V'(0,4)$, $V(10,4)$;
$B'(5,0)$, $B(5,8)$

15. $\dfrac{(x-1)^2}{4} + \dfrac{(y+2)^2}{16} = 1$; center $1, -2)$; $F'(1, -2 - 2\sqrt{3})$,
$F(1, -2 + 2\sqrt{3})$; $V'(1, -6)$, $V(1, 2)$; $B'(-1, -2)$, $B(3, -2)$

17. $\dfrac{(x+\frac{1}{2})^2}{8} + \dfrac{(y+\frac{3}{2})^2}{4} = 1$; center $(-\frac{1}{2}, -\frac{3}{2})$; $F'(-\frac{5}{2}, -\frac{3}{2})$, $F(\frac{3}{2}, -\frac{3}{2})$;
$V'(-\frac{1}{2} - 2\sqrt{2}, -\frac{3}{2})$, $V(-\frac{1}{2} + 2\sqrt{2}, -\frac{3}{2})$; $B'(-\frac{1}{2}, -\frac{7}{2})$, $B(-\frac{1}{2}, \frac{1}{2})$

19. $\dfrac{(x+2)^2}{289} + \dfrac{(y+3)^2}{225} = 1$; center $(-2, -3)$; $F'(-10, -3)$, $F(6, -3)$;
$V'(-19, -3)$, $V(15, -3)$; $B'(-2, -18)$, $B(-2, 12)$

21. $\dfrac{x^2}{9} + \dfrac{y^2}{4} = 1$.   23. $\dfrac{x^2}{41} + \dfrac{y^2}{16} = 1$ 25. $\dfrac{x^2}{25} + \dfrac{y^2}{4} = 1$

27. $\dfrac{x^2}{25} + \dfrac{(y-2)^2}{9} = 1$ 29. $\dfrac{(x-5)^2}{64} + \dfrac{(y-4)^2}{25} = 1$ 31. $x^2 + 4y^2 = 49$

33. 94.6 and 91.4 million miles   35. $\dfrac{20\sqrt{5}\,\text{ft}}{3} = 14.9\,\text{ft}$

37. $\dfrac{(x-2)^2}{40} + \dfrac{y^2}{49} = 1$ 39. $5x^2 + 9y^2 = 180$ 45. $\dfrac{x^2}{36+p} + \dfrac{y^2}{p} = 1,\ p > 0,$

$\dfrac{x^2}{81} + \dfrac{y^2}{45} = 1$

Section 10, *page* 101

1. $(\pm 5, 0)$, $(\pm\sqrt{34}, 0)$, $\tfrac{18}{5}$, $3x \pm 5y = 0$
3. $(\pm 4, 0)$, $(\pm 2\sqrt{5}, 0)$, 2, $x \pm 2y = 0$
5. $(0, \pm 7)$, $(0, \pm 7\sqrt{2})$, 14, $y \pm x = 0$
7. $(2, \pm 6)$, $(2, \pm 10)$, $\tfrac{64}{3}$, $4y \pm (3x - 6) = 0$
9. $(-3 \pm 4,\ 2)$, $(-3 \pm 5,\ 2)$, $\tfrac{9}{2}$, $3(x + 3) \pm 4(y - 2) = 0$

11. $\dfrac{(x-2)^2}{4} - \dfrac{(y-2)^2}{9} = 1$; center $(2, 2)$; $V'(0, 2)$, $V(4, 2)$; $F(2 \pm \sqrt{13}, 2)$

13. $\dfrac{(x-5)^2}{4} - \dfrac{y^2}{9} = 1$; center $(5, 0)$; $V'(3, 0)$, $V(7, 0)$, $F(5 \pm \sqrt{13}, 0)$

15. $\dfrac{(y-1)^2}{4} - \dfrac{(x-6)^2}{49} = 1$; center $(6, 1)$; $V'(6, -1)$, $V(6, 3)$; $F(6, 1 \pm \sqrt{53})$

17. $\dfrac{x^2}{9} - \dfrac{y^2}{16} = 1$ 19. $16y^2 - 5x^2 = 19$ 21. $\dfrac{(x-1)^2}{9} - \dfrac{(y+2)^2}{25} = 1$

23. $\dfrac{x^2}{25} - \dfrac{(y-6)^2}{36} = 1$ 29. $\dfrac{x^2}{4} - \dfrac{y^2}{12} = 1$

31. $\dfrac{x^2}{a^2} - \dfrac{y^2}{16-a^2} = 1,\ 16 - a^2 > 0;\ \dfrac{x^2}{4} - \dfrac{y^2}{12} = 1$

Chapter 4

Section 1, *page* 107

1. $3x' + 2y' = 0$
3. $y'^2 = 6x'$
5. $3x'^2 + 4y'^2 = 8$
7. $4y'^2 - 5x'^2 = 20$
9. $x'y' = 11$
11. $x'^3 - y' = 0$
13. $(4, 2)$, $x'y' = 12$
15. $(2, 0)$, $2x'^2 + 2y'^2 = 3$

17. $(-4, -2)$, $3x'^2 - 2y'^2 = 6$ 19. $(4, -1)$, $x'^2 - x'y' + y'^2 = 48$

21. $(-2, 2)$, $x'^3 + 2x'y' - x'^2 = 0$ 23. $y'^2 + 4x' = 0$

25. $y'^2 + 10x' = 0$ 27. $2x'^2 = 7y'$.

Section 2, page 111

1. $y' + 2 = 0$ 3. $x'^2 - y'^2 = 8$ 5. $3x'^2 + y'^2 = 2$

7. $y'^2 - 4x' = 0$ 9. $45°$ 11. $22\frac{1}{2}°$

Section 3, page 117

1. A rotation through $45°$ followed by a translation yields $y''^2 - 4x'' = 0$.

3. A rotation through arctan $\frac{3}{4}$ followed by a translation gives $x''^2 - y''^2 = 4$.

5. A rotation through arctan $\frac{2}{3}$ and a translation yields $2x''^2 + 3y''^2 = 6$.

7. A rotation through arctan $\frac{2}{5}$ and a translation yields $4x''^2 + y''^2 = 4$.

Section 5, page 122

1. Ellipse   3. Hyperbola   5. Hyperbola

7. Ellipse   9. The lines $3x - 2y + 1 = 0$ and $2x + y = 0$

11. The lines $x + y - 2 = 0$ and $x + y = 0$

13. The line $2x - y + 1 = 0$

15. The point $(-1, -\frac{3}{2})$ 17. Ellipse   19. No graph

21. The lines $x' = 0$ and $y' = 0$ 23. Hyperbola

Chapter 5

Section 1, page 127

1. $\dfrac{\pi}{3}, \dfrac{3\pi}{4}, \dfrac{5\pi}{6}$ 3. $\dfrac{5\pi}{4}, \dfrac{7\pi}{4}, \dfrac{11\pi}{6}$ 5. $20°, 135°, 52.5°$

11. a) $(-3, 240°), (-3, -120°), (3, -300°)$
   b) $(6, 330°), (-6, 150°), (-6, 210°)$
   c) $(-2, 0°), (2, -180°), (-2, 360°)$
   d) $(4, -345°), (-4, 195°), (-4, -165°)$

Section 2, page 131

3. $(3, 3)$ 5. $(0, 0)$ 7. $(-\frac{1}{2}, \frac{1}{2}\sqrt{3})$

9. $(-4, -4)$ 11. $(3, 90°)$ 13. $(0, \theta)$ where $\theta \geqslant 0°$

15. $(5, 270°)$ 17. $(12, 330°)$

19. $(5, 307°)$, nearest degree   21. $(13, 67°)$

23. $r \cos \theta = 3$ 25. $r = \dfrac{3}{2 \cos \theta - \sin \theta}$

27. $r \sin^2 \theta = 4 \cos \theta$ 29. $r = 4$ 31. $r \sin^2 \theta - 9 \cos \theta = 0$

33. $r = 2 \cos \theta$ 35. $x^2 + y^2 = 16$ 37. $y = \sqrt{3} x$

39. $y = 6$ 41. $x^2 + y^2 - 8y = 0$ 43. $2xy = a^2$

45. $3(x+1)^2 + 4y^2 = 12$ 47. $4x + 3y = 2$ 49. $x - 3y = 1$

Section 6, page 148

11. $r \sin \theta = -2$ 13. $r \cos \theta = -3$

15. $r \cos (\theta - 225°) = 4$ 17. $(3, 90°)$, 3

19. $(-2, 0°)$, 2   21. $(-\frac{7}{2}, 90°)$, $\frac{7}{2}$

23. $r - 8 \cos \theta = 0$ 25. $r + 10 \cos \theta = 0$

27. $r^2 - 8r \cos \theta + 12 = 0$ 29. $r - 10 \cos (\theta - 45°) = 0$

31. $r^2 - 6 \cos (\theta - 120°) + 5 = 0$

Section 8, page 158

1. $(1, 60°)$, $(1, 300°)$ 3. $(3\sqrt{2}, 45°)$, $(3\sqrt{2}, 315°)$

5. $(a, 90°)$, $(a, 270°)$ 7. $(2, 0°)$

9. $(\frac{4}{3}, 60°)$, $(\frac{4}{3}, 300°)$

11. $(1, 0°)$, $(-\frac{3}{5}, 233°)$. $\text{Arcsin} (-\frac{4}{5}) = 233°$, to the nearest degree

13. $(0, 90°)$, $(2\sqrt{3}, 30°)$, $(-2\sqrt{3}, 150°)$ 15. $(1, 60°)$, $(1, 300°)$

17. $(3, 180°)$ 19. $(2, 60°)$, $(2, 300°)$, $(-1, 180°)$

Chapter 6

Section 3, page 166

5. $x^2 + y^2 = 25$ 7. $9x^2 + 16y^2 = 144$

9. $y^2 = -2(x - 1)$ 11. $16(x+4)^2 + 9(y-3)^2 = 144$

13. $x^2 + y^2 = 2x$

15. The part of the line $3x + 2y = 6$ joining $(2, 0)$ and $(0, 3)$

17. The part of the line $x + y = 5$ joining $(1, 4)$ and $(4, 1)$

19. The part of the parabola $y = -8(x - 2)$ for which $0 \leqslant x \leqslant 2$

21. $x = t^{-1}$; $y = t + 2$ 23. $x = a \sin^4 \theta$; $y = a \cos^4 \theta$

Section 5, page 170

1. $x = 40\sqrt{2}t, y = 40\sqrt{2}t - 16t^2$; $(x - 100)^2 = -200(y - 50)$. The greatest height is 50 ft, and the ball strikes the ground 200 ft away.

3. $x = v_0 t$, $y = -16t^2$; $16x^2 = -v_0^2 y$. In 2 seconds the projectile falls 64 ft, and travels $2v_0$ ft horizontally.

7. $x = 4\pi t - 2t \sin \pi t, y = 4 - 2t \cos \pi t$

Chapter 7

Section 5, page 184

21. $(3, 2)$

23. $(0, 0), (2\sqrt[3]{12}, 2\sqrt[3]{18})$

25. $(-1, 3), \left(\dfrac{1 \pm \sqrt{17}}{2}, \dfrac{9 \pm \sqrt{17}}{2}\right)$

27. $(\pm 3, 2), (\pm 3, -2)$

Chapter 8

Section 3, page 193

11. 11

13. $2\sqrt{26}$

15. $(3, 0, 2), (3, 2, 0), (0, 2, 2), (3, 2, 2)$

17. The yz plane.

19. The xy plane.

21. The plane 5 units below the xy plane.

23. The plane parallel to the y axis and passing through $(0, 0, 3)$ and $(2, 0, 0)$.

25. The plane parallel to the z axis and passing through $(3, 0, 0)$ and $(0, 5, 0)$.

27. The right circular cylinder of radius 3 whose axis is the x axis.

29. The right circular cylinder of radius 3 with axis parallel to the x axis and passing through $(0, 3, 0)$.

31. The right circular cylinder of radius 4 with axis parallel to the z axis and passing through $(4, 0, 0)$.

33. The plane determined by the points $(2, 0, 0)$, $(0, 4, 0)$ and $(0, 0, 2)$.

Section 6, page 207

1. $C(0, 2, 0)$, r = 2

3. $C(2, -4, 3)$, r = 6

5. $\dfrac{x^2}{a^2} + \dfrac{y^2}{b^2} + \dfrac{z^2}{a^2} = 1$

Chapter 9

Section 3, page 219

1. $-2\mathbf{i} + 8\mathbf{j}, 6\mathbf{i} - 2\mathbf{j}$ 3. $2\mathbf{i} - 6\mathbf{j}, 4\mathbf{i} + 2\mathbf{j}$ 5. $\frac{3}{5}\mathbf{i} + \frac{4}{5}\mathbf{j}$

7. $\dfrac{1}{\sqrt{17}}\mathbf{i} - \dfrac{4}{\sqrt{17}}\mathbf{j}$ 9. $\dfrac{2}{\sqrt{13}}\mathbf{i} - \dfrac{3}{\sqrt{13}}\mathbf{j}$ 11. $(4, -1)$

13. $(\frac{3}{2}, -1)$ 17. $(-3, 4), (0, 5)$ 19. $(-\frac{4}{3}, -1), (\frac{1}{3}, 3)$

Section 4, page 222

1. $8\mathbf{i} - 2\mathbf{j} + 3\mathbf{k}, -2\mathbf{i} + 6\mathbf{j} + 5\mathbf{k}$

3. $-6\mathbf{i} + 2\mathbf{j} + 4\mathbf{k}, 2\mathbf{i} + 4\mathbf{j} + 4\mathbf{k}$ 5. $\frac{3}{7}\mathbf{i} + \frac{2}{7}\mathbf{j} + \frac{6}{7}\mathbf{k}$

7. $\frac{2}{3}\mathbf{i} + \frac{2}{3}\mathbf{j} + \frac{1}{3}\mathbf{k}$ 9. $\dfrac{2}{\sqrt{14}}\mathbf{i} + \dfrac{1}{\sqrt{14}}\mathbf{j} + \dfrac{3}{\sqrt{14}}\mathbf{k}$

11. $5\mathbf{i} + 4\mathbf{j} - 3\mathbf{k}$ 13. $4\mathbf{i} + 5\mathbf{j} + 8\mathbf{k}$

15. $3\mathbf{i} - \mathbf{j} + 4\mathbf{k}, (3, -1, 4); 5\mathbf{i} + \mathbf{j} + \mathbf{k}, (5, 1, 1)$

17. $\frac{7}{2}\mathbf{i} + \frac{9}{2}\mathbf{j} + \frac{13}{2}\mathbf{k}, (\frac{7}{2}, \frac{9}{2}, \frac{13}{2}); 3\mathbf{i} + 4\mathbf{j} + 6\mathbf{k}, (3, 4, 6); \frac{5}{2}\mathbf{i} + \frac{7}{2}\mathbf{j} + \frac{11}{2}\mathbf{k}, (\frac{5}{2}, \frac{7}{2}, \frac{11}{2})$

19. $(-5, -2, -6), (7, 7, 12)$

Section 5, page 227

1. $-2, -\frac{2}{27}$ 3. $13, \frac{13}{135}$

5. $\frac{19}{3}\sqrt{3}, \frac{19}{3}(\mathbf{i} - \mathbf{j} - \mathbf{k})$ 7. $55°$

9. $64°, 90°, 26°$

Section 6, page 232

1. $3x - 2y + 5z - 4 = 0$ 3. $2x - 3y - 4z = 8$

5. $3x + 9y - 4z = 0$ 7. $2x + 3y - 4z = 2$

9. $\frac{1}{6}\sqrt{6}$ 11. 0 15. $\frac{8}{3}$ 17. $\frac{2}{7}\sqrt{21}$

Section 8, page 239

1. $\dfrac{x - 4}{-2} = \dfrac{y + 3}{3} = \dfrac{z - 5}{4}; x = 4 - 2t, y = -3 + 3t, z = 5 + 4t$

3. $\dfrac{x - 1}{2} = \dfrac{y - 1}{3} = \dfrac{z - 2}{-1}; x = 1 + 2t, y = 1 + 3t, z = 2 - t$

5. $\dfrac{x - 2}{2} = \dfrac{y + 1}{1}, z - 1 = 0; x = 2 + 2t, y = -1 + t, z = 1$

7. $x = 0, z = 0; x = 0, y = t, z = 0$

9. $\dfrac{x-1}{-3} = \dfrac{y-2}{2} = \dfrac{z-3}{-3}$ 11. $\dfrac{x-1}{-1} = \dfrac{y}{2} = \dfrac{z-2}{-1}$

13. $\dfrac{y-5}{1} = \dfrac{z-4}{1}, x = 2$ 15. $y - z = 0, z - 4 = 0$

17. $\dfrac{x-2}{3} = \dfrac{y-3}{-1} = \dfrac{z}{2}$ 19. $\dfrac{x}{2} = \dfrac{y-3}{-3} = \dfrac{z-6}{1}$

21. $60°$ or $120°$ 23. $\frac{4}{9}, \frac{1}{9}, \frac{8}{9}$ 25. $\frac{2}{5}, \frac{2}{15}, -\frac{11}{15}$

27. $\dfrac{1}{\sqrt{35}}, \dfrac{3}{\sqrt{35}}, \dfrac{5}{\sqrt{35}}$ 29. $\cos\theta = \frac{2}{3}$

31. The yz plane at $(0, -5, -12)$, the xz plane at $(10, 0, 3)$, the yx plane at $(8, -1, 0)$.

33. The yz plane at $(0, 1, 2)$, the xz plane at $(3, 0, 4)$, the xy plane at $(-3, 2, 0)$.

Section 10, page 246

1. $4\mathbf{i} + 4\mathbf{j} - 2\mathbf{k}; \frac{2}{3}\mathbf{i} + \frac{2}{3}\mathbf{j} - \frac{1}{3}\mathbf{k}$ 3. $2\mathbf{j} - 2\mathbf{k}; \dfrac{\sqrt{2}}{2}\mathbf{i} - \dfrac{\sqrt{2}}{2}\mathbf{k}$

5. $25\mathbf{k}, \mathbf{k}$ 7. 5 9. $7\sqrt{6}$ 11. $\frac{2}{9}\sqrt{3}$

13. $\frac{1}{2}$ 17. $9x + 7y - z = 25$

19. $\frac{274}{781}\sqrt{781}$ 21. $\dfrac{x}{3} = \dfrac{y}{2} = \dfrac{z}{4}$

23. $3x - 7y + 8z = 4$ 25. $\frac{1}{3}\sqrt{170}$

27. $\dfrac{\sqrt{42}}{14}$ 31. $\dfrac{43\sqrt{227}}{227}$

33. The lines intersect.

Chapter 10

Section 2, page 253

We assume that the data in the exercises of Chapter 10 justify the retention of three significant figures in the answers.

1. $y = -0.718x + 8.71$

3. $y = 0.100x + 10.0$

5. $N = 1380t + 6300$

Section 3, page 257

1. $y = 0.121x^{3.13}$

3. $t = 0.259s^{0.488}$

5. $p = 103v^{-1.39}$

Section 4, page 260

1. $y = 2.00 \cdot 10^{0.131x}$

3. $y = 9.91 \log x + 3.03$

5. $T = 99.5 \cdot 10^{-0.0989t}$

7. $V = 3.87 \log P - 0.828$

Index

Index